Social Scientists of the Islamic Era

Volume 3 of the 8-volume series on

Scientists of the Islamic Era

"Scientist of the Islamic Era" is a book series encompassing eight volumes; the present book is volume 3 titled "Social Scientists of the Islamic Era". It covers 74 Social Scientists encompassing philosophers, historians, physical geographers, and qadhis, as well as the conventional sociology, political science, management sciences, economics, business, trade, anthropology, and linguists including letters. The period of coverage is part 1 of the Islamic Era, from AD 610 to 1400. They commanded exceptional breadth in their learning and deepest insights in their specializations; they greatly strengthened the foundations and expanded the frontiers of the fields of knowledge in the Social Sciences.

Each scientist is briefly described. First, the name of the scientist is disambiguated, and an attempt is made to correct the misrepresentations common in the European translations. Salient scientific contributions of each scientist are briefly highlighted, a difficult task because of the fact that many of these scientists were polymaths. For each scientist we have provided a biographical summary to help picture their love and craving for knowledge, and the motivations and opportunities for them to do their research.

It is our objective that this third volume in the series will inform the Muslims about the wealth of their scientific heritage, and the next generations will feel inspired to surpass the excellence of their ancestors to enrich their heritage further, and be, like their ancestors, the flag bearers of world civilization in the social sciences.

Muslims are now excelling in research with superb agility, and this series on Scientists of the Islamic Era will further stimulate this Renaissance in the Muslim world.

Social Scientists of the Islamic Era

Volume 3 of the 8-volume series on

Scientists of the Islamic Era

Abdur Rahim Choudhary, Ph.D.

Muslim Voice

MV Publishers

Published by MV Publishers, a subsidiary of Muslim Voice,
12719 Hillmeade Station Dr, Bowie, MD 20720, USA.
MVPublishers@muslimvoice.org

ISBN 978-1-956601-20-6

First edition 2024

United States of America

Choudhary, Abdur Rahim, 1944–
8-Volume Series on Scientists of the Islamic Era,
Volume 3, Social Scientists of the Islamic Era

Muslim Voice

ISBN 978-1-956601-20-6

To the Muslim Ummah

Content

Preface to the 8-volume series on Scientists of the Islamic Era (Updated)

For a period of more than a millennium, Muslim Scientists have done foundational research in all scientific disciplines, and also greatly expanded the frontiers of science. However, our people often do not have a clear idea about our scientific heritage. We decided to write a series of books on "*Scientists of the Islamic Era*" that would be readily available to our generation and the coming generations, and provide motivation for excellence in the world civilizational dialogue, as well as to know our religious inspiration for scientific research and progression.

The young generations, especially those in Europe and Americas, have now opened their hearts and minds with a renewed desire for the truth about Islam and Muslims, being less influenced by historical biases and religious prejudices. The eight books in the series on *Scientists of the Islamic Era* seek to serve their youthful thirst for the truth.

Another reason for this series on "*Scientists of the Islamic Era*" is to produce a consciousness among the present-day academicians and scientists about the foundational contributions that the Muslim scientists made to all scientific disciplines, as well as how they expanded the frontiers of these disciplines. This fact is evidenced in the books in this series. However, this fact is not widely known because the present-day literature does not reference these original sources. The chain of scholarly references ends in European

Renaissance, with occasional references to Greek scientists, but bypassing the millennium worth of research by Muslim scientists, who established the foundational principles and greatly expanded the frontiers of science.

In addition, the work seeks to fill a void, as no such series of books currently exists.

Islamic Era constitutes the period from 610 AD, when the Prophet received his first revelation, to 1922 AD, when the Ottoman Caliphate ended and the Turkish Republic began. We have divided the period in two parts: part 1 from 610 to 1400, and part 2 from 1400 to1922. The era is divided at an epoch when much of the works by the Muslim Scientists had already been translated into European languages, had become widely available, and had begun to produce Renaissance in Europe.

Each of the two parts of the Islamic Era is covered by the following four volumes, eight volumes in all.

Volume 1 is for Natural Sciences that include mathematics, astronomy, cryptoanalysis, chemistry, cartography, physics, and engineering based on these disciplines such as mechanics, automation, and robotics.

Volume 2 is for the Medical Sciences that include physicians, nurses, surgeons, herbalists, medical researchers, and medical writers.

Volume 3 is for the Social Sciences that include philosophers, historians, physical geographers, qadhis (judges), as well as the conventional sociology, political science, management sciences, economics, business, trade, anthropology, and linguists including letters.

Volume 4 is for the Religious Sciences that include analogists, Mohaddessin (historical fact checkers), jurisprudents, mofassarin (Quranic meanings and interpretations), and spiritualists (sciences of the paths (tariqas)).

Categorizing fields of scientific enquiry is currently done in the limited context of the European experience. For example, there is no name for the Spiritualist people like Sufis; Muhaddis who compile events of historic significance and fact check them for validity and accuracy; and the Faqih who exercise jurisprudential judgement using analogies and scriptural reasoning. In an attempt to classify the fields of research worldwide, it is necessary to use a classification scheme that is global and encompassing, rather than restrictive to a region and thus limiting.

It therefore becomes necessary to introduce English terminology for large and overarching fields of study that correspond to specializations like Hadith, Fiqh, and Religious Studies in Islam. Corresponding terms for the Christian and Jewish religions exist in great detail incorporating their taxonomies to small granularities. For example, there are a large number of terms introduced to classify Catholic hierarchy within the Church. While the categories of people who study and practice Islam is vast, no terminology to describe them has unto now been ascribed. Reasons for the circumstance are myriads but it is now time to remove the gaps, and enlarge the research and make it global and free from omissions and restrictions. We have, therefore, introduced new terms for scientific studies of these topics. A minimal number of new terms are coined for English language because no appropriate terms currently exist to represent the corresponding meanings in Arabic.

Factologists: The term represents the scientists who investigate criteria used for the verification and validation of orally transmitted or written historical facts. These were developed chiefly by the Mohaddessin. The science, among other things, uses what is known as 'knowledge about the people (Ilm-ur-Rijal)' encompassing all people who narrated any traditions connected with the Prophet. This domain of investigation did not previously exist and is exclusively developed from scratch by the Mohaddessin. No such scientific research domain exists in European social sciences, hence a lack of terminology for it. Europeans did not learn this science. They have, therefore, not employed it in their study of disciplines like history, or in the modern age to weed out spurious phenomena like the fake news or alternate facts.

Analogists: The term represents scientists who investigate judicial inferences and precedence based on the analysis of scriptures or constitution. It is developed chiefly by the people of the Fiqh. They employ scripture and compiled traditions incorporating precedence to decide judiciously. They use direct analogies with the material in precedences. Analogists do not use intellectual interpolation and extrapolation; such a use would turn the analogist into qiasist. This science is somewhat similar to the practice of judicial precedence in constitutional judicial systems. There is, however, an important difference, namely, only those occurrences of precedence are incarporated that happened in early Islamic era and which have been verified and validated by the factologists.

Qiasist: The term is used and developed chiefly by the people of the Fiqh. Like analogists, scripture, compiled traditions, and precedence are used. Application of analogy solves most of the judicial cases. However, in some judicial cases the application of the analogy does not lead to a decision. It can be for many reasons. First, there may exist no analogy because the world conditions have changed in terms of the possible set of actions and events. An example would be the performance of rituals like prayer, fasting, and hajj during air travel. Second, the application of analogy may not yield a clear and unique judgement because the set of analogies available in a judicial case and related circumstances may be varied and diverse. In general, when the application of analogy is insufficient to arrive at a judicial decision the Qadhi (judge) needs to resort to the use of his own intellect to arrive at a Qias (guess) with respect to what the correct judgement should be in a particular situation. Qias uses direct analogies and minimally employs intellectual interpolation and extrapolation. This science is somewhat similar to the practice of judicial precedence in constitutional judicial systems, except that only those occurrences of precedence are incorporated that happened in early Islamic era and which have been verified and validated by the factologists.

The difference between analogy and qias is that the intellect of the judge is used in qias while it is not used in analogy. Since intellect is used in all human actions, and Muslims themselves may not be unanimous about it, some clarification is needed. In the case of a clear analogy the use of intellect is mostly not specific to the particular qadhi since most qadhis would be onboard. However, that is generally not be the case in the case of qias so that the judgement may be more specific to the particular qadhi. Such distinction and finetuning points to the sophistication of Islamic judicial system. No such distinction and corresponding sophistication exists regarding the use of precedence among the European justice systems, and that can lead to arbitrariness and systematic bias.

Awwalagists: The term represents scientists who perform analysis to bring the textual meanings back to their roots, as the Mufassarin have done for the meanings of Quran, which science is known as Tafsir and Ta'wil in Arabic. This is not the same as exegetes because the term exegesis applies heavily to the Bible paying little heed to the process for other scriptures and languages. Further, the term exegesis is also used in broad connotations such as politics and literature. Therefore, the term awwalagists is coined to represent the concept of Ta'wil and Tafsir.

Awwalagist carries a meaning that is fresh even in Muslim tradition. It is distinct from the term mufassir which carries quite a historical baggage and is entrenched in sectarian controversy. Independent of such historical baggage and controversy, the term awwalagist carries a spirit of research in pursuit of the truth, irrespective of who the researcher is in the conventional sectarian landscape.

Muhaddis: This science is chiefly an application of factology. A muhaddis collects traditions by seeking out people who have a tradition to narrate. The narrator is known as the Ravi. It is incumbent upon a ravi to satisfactorily explain how he or she received the narration; starting from the source and declaring each and every intermediary ravi between the source and himself or herself. The factologist uses a most detailed and expansive catalogue of all the ravis that exist in the total landscape of narrations. A detailed biography is compiled about each ravi and it includes factors like piety, truthfulness, profession, etc. Any doubtful character among the chain of ravis for a narration will make the narration fail the test of factology. A great body of knowledge exists in the science of factology, for example the evaluation of the piety of each ravi by reviewers over the history. History thus is not just a collection of assertions by historians, rather each such assertion is fact checked most thoroughly according to most strict and well specified criteria. Therefore, Muslim muhaddis are historians with far greater reliability than the European historians who can maneuver with respect to accuracy without rigorous fact checks applied to their works.

Mujtahid: A mujtahid is someone who can use ijtihad, which roughly means exercising qias in a mostly error-free way. A mujtahid is knowledgeable about the scriptures, their tafsir and tawil, the application of analogies, and limitations of qias, etc. He or she is also an established analogist, qiasist, factologist, muhaddis, and awwalagist, etc.

A mujtahid is one who can exercise ijtihad, which is a process that exercises all the above-mentioned elements. One distinguishing element of ijtihad is the ability to practice qias that is widely regarded as error free.

Muslims regard such a person as an Imam, meaning a religious and sometimes even a political leader whom they follow in matters of guidance and inspiration. Five such Imams are well known: Imam Abu Hanifa, Imam Malik, Imam Shafi, Imam Hanbal, and Imam Jaffer Sadiq. Their teachings are the same except possibly in those cases where they employ qias, in which situations they can differ.

Fiqh and Faqih: A mujtahid specifies the rules of judicial decision making. This set of rules is called a Fiqh. Different fiqhs are the same, except possibly where the mujtahid makes recourse to qias. A faqih is one who can use fiqh to make judicial decisions. A qadhi is a faqih who is officially appointed to make binding judicial decisions that are enforced under the state authority.

Ruhaniyat: The term represents the science of Islamic seeking of the truth about life, based on Salat (prayer), Saum (fasting) and Zikr (recall) in general. These people are Scientists of Ruhaniyat. Non-Muslim investigators like the Christian Saints might also be included if they meet the criteria, even though Muslims enjoy far greater freedom from the central authority (which does not exist among Muslims), and have made ruhaniyat into a genuine science with detailed methodologies and verification markers along the path.

English terms for some of the other sciences are retained for the Scientists of the Islamic Era as well, because such terms are approximate enough; such as mathematics, physics, surgeon, sociology, economics, political science, poetry, philology, etc. Even though Muslims approach these topics with subtly different angles, for now we will decide to use these terms in the context of Muslim Scientists as well. However, a need still exists to introduce additional terms in order to represent the research of the scientists of the Islamic Era at a finer resolution. We will leave that for a future edition.

It is also necessary to remark how the historical periods are currently named and defined. These are currently European centric; as well as confusing and lacking clear rational to the terminology. We have corrected the anomaly at a small scale, namely as it concerns the definition of the "Middle Ages". These are also variously called Dark Ages or Medieval Period (also written as Mediaeval Period). The period has vague start and stop durations, as treated by different historians. We have, therefore, redefined it crisply, and also renamed it. Instead of starting it from 476 AD, a vague reference point to the fall of Roman Empire, we start it at 610 AD, a reference point to the beginning of the Islamic Era when the Qur'anic revelation began. We end this Islamic Era with the fall of Ottoman Caliphate in 1922. Thus, the Middle Ages is largely replaced with the Islamic Era, and it extends into what is commonly and non-transparently called the Modern Era, Modern Age or Modern Period. Doing so, makes the scheme less European centric and more logically consistent.

We present this series of eight volumes to the readers to share with them the wealth of scientific excellence that these scientists contributed towards greatly extending the frontiers of all these sciences. In particular, they enhanced the world civilization towards prosperity, democracy, justice and peace; as well as produced Renaissance in Europe. The series also aims to bring awareness to the Muslim readers about their role as the torch bearers of science and civilization, and to serve the upwelling thirst that the young generation have for the truth about Islamic civilization. The series urges the academicians and researchers of the world, especially the Europeans and Americans, to learn and celebrate the Muslim giants of science upon whose shoulders they stand, and without whom the present-day scientific achievements could not have been possible.

Researchers like Professor Fuat Sezgin have, at great length, researched and investigated contributions of the Muslim scientists. He has edited 1600 volumes. Such work is invaluable for projects like ours. Some of this work is summarized online, such as in Wikipedia articles available under GNU free document license.

We aim to serve our community, inspire them and our coming generations, and inform them of their role as torch bearers of excellence in world science, technology, and civilization.

The Muslim scientists lived an integrated life with no conflict between their religion and their scientific passions; and, unlike the many present-day scientists, a question never occurred that their scientific passion somehow needed to be separate from their religious inspirations. This is also obvious from the fact that most scientists were themselves experts in Islamic jurisprudence, hadith and Quran. In reality, their scientific work was also their religious worship because Islam showed them the necessity to do science and motivated them for it. Islam encouraged their scientific passion by equating it with religious worship. No wonder they achieved scientific excellence with amazing integrity, generosity and grace.

The Muslim education system was very different from what the world has today; and judging from the results, it was greatly more affective and better integrated into overall life. The biographical summaries will provide tiny glimpses into this system. It avoided narrow specializations and produced polymaths. We have Muslim scientists who are simultaneously excellent Quranic scholars, jurisprudents, mathematicians, astronomers, physicists, medical professionals, chemists, botanists, physiologists, poets, men of letters, and grammarians, etc.

To keep this series on "Scientists of the Islamic Era" brief we have included each scientist only once, under a category we deemed appropriate, given their works and biography. As a consequence, for instance, not all natural scientists appear in volume 1 which is dedicated to the Natural Scientists, because we decided to include some in Volume 2 which is dedicated to the Medical Scientists; similarly, not all medical scientists appear in Volume 2 because we decided to include some in Volume 1. This situation invariably occurs across the board. Another consequence is that, for instance, if a scientist is included in Volume 2 for the Medical Scientists, his or her achievements in Astronomy, for instance, are not highlighted.

For such reasons, the preface to the series will get edited as the series progresses and situations for it arise.

The scientists are listed in chronological order, allowing an opportunity to correlate scientific tides and ebbs with political and religious ups and downs.

They could have been ordered according to the significance of their scientific contributions; that, however, is problematic because it is difficult, if not impossible, to assess the importance of research and compare across different scientific disciplines within sciences.

The order could have been sequenced according to how well the scientists are known today; that too is problematic because not all excellent scientists are well-known today, and, those who are, generally are made famous by the European commentators, who often did not know their works in original Arabic, and did not reflect the actual significance of their research. The well-known-ness is fairly arbitrary. For instance, Omar Khayyam is celebrated today for his Rubaiyat, which was something he did on the side, while his real works were in mathematics, a fact that is largely obscured.

This series of books should add to the impulse that is now thrusting the Muslims into the world of science and technology with increasing excellence in their achievements, signaling that their own Renaissance has now renewed.

Muslim Voice
Bowie, MD, USA.
July 29th, 2024.

Preface to the First Edition of Volume 3: Social Scientists

"Scientist of the Islamic Era" is a book series encompassing eight volumes. The present book is volume 3 titled "Social Scientists of the Islamic Era" that covers 74 philosophers, historians, physical geographers, and qadhis, as well as the conventional sociology, political science, management sciences, economics, business, trade, anthropology, and linguists including letters. The period of coverage is part 1 of the Islamic Era, from AD 610 to 1400.

Most social scientists in this book are multidisciplinary and interdisciplinary; and they also excelled in jurisprudence, hadith, philology, and poetry. They commanded exceptional breadth in their learning and deepest insights in their specializations; and, thus, greatly strengthened the foundations and expanded the frontiers of fields of knowledge.

It is our objective that this third volume in the series will inform the Muslims about the wealth of their scientific heritage, and the next generations will feel inspired to surpass the excellence of their ancestors to enrich their heritage further, and be, like their ancestors, the flag bearers of world civilization regarding the social sciences. Our objective is also for the academic community to learn the truth about how science grew by leaps and bounds during the Islamic era. The book series shall quench the thirst of the youth, especially in Europe and Americas, to discover the truth about Muslim contributions to the world science and civilization.

Muslims are now excelling in scientific research and civilizational leadership with superb agility. The books in the series on "Scientists of the Islamic Era" are expected to add impetus to this Renaissance in the Muslim world.

Abdur Rahim Choudhary, Ph.D.
Bowie, Maryland, USA
arc@muslimvoice.org
July 29th, 2024.

Social Scientists

Social Sciences community in Islamic Era was dominated by the Muslim scientists; the European scientists during this time were virtually nonexistent, owing to Europe being in the "Dark Age". When they started to emerge a little before the European Renaissance, they did so based on the research works of the Muslim scientists done for the prior seven centuries, which had already been translated into European languages, and had become broadly available.

These facts are obvious even if one examines not the entire scientific works by the Muslim scientists but only a subset of those that had been very visibly translated into European languages.

This book describes 74 social scientists from part 1 (610-1400) of the Islamic era covering the disciplines of philosophers, historians, physical geographers, qadhis, as well as sociology, political science, management sciences, economics, business, trade, anthropology, and linguists including letters. Each scientist is briefly described. First, the name of the scientist is disambiguated, and an attempt is made to correct the misrepresentations all too common in the European translations. Salient scientific contributions of each scientist are briefly highlighted, a difficult task because of the fact that most of these scientists were polymaths. For each scientist we have provided a biographical summary to help picture their love and craving for knowledge and the motivations and opportunities available for them to do their research.

The list of 74 social scientists, that are covered in this edition of the book, is given in the table below, in chronological order. Each entry in the table includes the year of death and a one-line description, including the name of the scientist, the time period in parenthesis, and the area(s) of specialization within the social sciences.

Table of 74 Social Scientists covered in this edition of the book.

708	Abd-Allah ibn Ibadh (d. 708, b. Basra), founder of Ibadhi sect
741	Ibn Shihab al-Zuhri (670–741), historian
748	Wasil ibn Ata (700–748), theologian and founder of the Mutazilite school of Islamic thought
761	Ibn Ishaq (704–761), historian and hagiographer
774	Abu Mikhnaf (d. 774), historian
791	Al-Farahidi (c. 718 – 791), writer and philologist, compiled the first dictionary of the Arabic language, the Kitab al-Ayn
793	Al-Akhfash al-Akbar (d. 793, b. Basra), Arab grammarian
805	Muhammad al-Shaybani (749/50 – 805), father of international law
819	Hisham ibn al-Kalbi (d. 819), historian
821	Abu Amr al-Shaybani (d. 821/28), lexicographer and collector of Arabic poetry
835	Ibn Hisham (d. 835), historian and biographer
840	Al-Jarmi (d. 840), grammarian of Arabic Language
845	Ibn Sa'd (784–845), scholar and Arabian biographer
854	Khalifah ibn Khayyat (777–854), historian
870	Al-Zubayr ibn Bakkar (788–870), historian and genealogist
871	Ibn 'Abd al-Hakam (803–871), Egyptian historian
880	Fatima al-Fihri (800, Kaioruan – 880), science patron and founder of the Al Quaraouiyine mosque
892	Al-Baladhuri (820, Baghdad – 892, Baghdad), historian

898	Al-Mubarrad (826–898), grammarian and linguist
915	Al-Jubba'i (d. 915), philosopher and Mu'tazili theologian
934	Ibn Duraid (837, Basra – 934, Baghdad), geographer, genealogist, poet, and philologist
950	Al-Farabi (872-950), polymath, philosophy, physics, chemistry, Jurist
950	Ibn Fadlan (10th century), writer, traveler, political scientist, member of an embassy of the Caliph of Baghdad to the Volga Bulgars
956	Al-Masudi (896–956), historian, geographer, philosopher, traveler, (traveled to Spain, Russia, India, Sri Lanka and China, spent his last years in Syria and Egypt)
969	Ibn Hawqal (943–969), writer, geographer, and chronicler
977	Ibn al-Qūṭiyya (d. 977), Andalusian historian
991	Al-Maqdisi (946–991), geographer
995	Ibn al-Nadim (d. 995), encyclopedia catalogue of authors known as 'Kitāb al-Fihrist'
1030	Al-Musabbihi (977–1030), Fatimid historian
1050	Al-Mubashshir ibn Fatik (11th century), polymath, mathematical sciences, writer of biographies of sages.
1057	Al-Maʿarri (973–1057), blind philosopher, poet and writer
1169	Abu'l-Hasan al-Bayhaqi (1097–1169), historian and astronomer
1094	Al-Bakri (1014, Huelva – 1094, Cordoba), geographer and historian
1095	Al-Humaydī al-Azdi (1029–1095), historian
1144	Al-Jawaliqi (1074–1144), grammarian and philologist
1147	Ibn Bassam (1058, Santarem – 1147), poet and historian
1160	Ibn al-Qalanisi (c. 1071–1160), chronicler and historian
1066	Ibn Sidah (c.1007–1066), grammarian and lexicographer
1172	Ibn Zafar al Siqilli (1104–1172), polymath, Arab-Sicilian philosopher
1185	Al-Suhayli (1114–1185), grammarian and scholar of law.

1185	Ibn Tufail (1105–1185), polymath, Andalusian writer, novelist, philosopher, theologian, physician, astronomer, vizier, court official, political scientist
1187	Gerard of Cremona (1114–1187), Italian translator from Arabic to Latin.
1188	Usama ibn Munqidh (1095–1188), historian, politician, diplomat, political scientist
1196	Ibn Maḍāʾ (1116–1196), grammarian and mathematician
1198	Ibn Rushd (Averroes) (1126-1198) polymath, philosopher, mathematics, physics, medicine, astronomy, psychology, theology, jurisprudence, law, linguistics.
1201	Ibn al-Jawzi (1116–1201), heresiographer, historian, philologist, hagiographer
1217	Ibn Jubayr (1145–1217), geographer, traveler known for his detailed travel journals, poet
1222	Al-Jawbari (fl. 1222), writer, alchemist
1231	Abd al-Latif al-Baghdadi (1162, Baghdad – 1231, Baghdad), polymath, historian, Egyptologist, traveler, physician
1233	Ibn al-Athir (1160, Cizre – 1233, Mosul), historian, biographer
1235	Ibn Dihya (1150, Valencia – 1235, Cairo), philologist, grammarian, short biographies and poems from Andalusian and Moroccan poets, Muhaddis
1248	Al-Qifti (1172–1248), historian
1260	Ibn al-Abbar (1199, Valencia – 1260, Tunis), historian, poet, diplomat, political scientist, theologian
1262	Ibn al-Adim (1192, Aleppo – 1262, Egypt), biographer, historian
1270	Ibn Abi Usaibia (1203–1270), physician, historian, wrote 'Lives of the Physicians'
1274	Ibn Malik (1203/1204 or 1204/1025 – 21 February 1274), grammarian
1283	Zakariya al-Qazwini (d. 1283), proto-science fiction writer, physician, astronomer, geographer

1286	Ibn Sa'id al-Maghribi (1213–1286), geographer
1312	Ibn Manzur (1233–1312), lexicographer, linguist
1327	Al-Dimashqi (1256, Damascus – 1327, Safed), geographer
1333	Al-Nuwayri (1279–1333), historian, encyclopedist
1349	Hamdallah Mustawfi (1281–1349), geographer
1349	Shihab al-Umari (1300–1349), historian
1357	Ibn Juzayy (d. 1357), historian, poet
1368	Ibn Battuta (1304-1368), explorer, traveler
1374	Ibn al-Khatib (1313–1374), polymath, poet, writer, historian, philosopher, physician
1406	Ibn Khaldun (1332–1406), historian, sociologist, philosopher
1407	Ismail ibn al-Ahmar (1324–1407), historian
1429	Taqi al-Din Muhammad al-Fasi (1373–1429), historian, Maliki qadi
1442	Al-Maqrizi (1364–1442), historian
1450	Ibn Arabshah (1389, Damascus – 1450, Egypt), writer, traveler

We expect that this list will be expanded in subsequent editions, as further research is carried out.

A brief description for each scientist is provided, each in a separate subchapter. The 74 subchapters, that follow, are each dedicated to a single social scientist. Some chapters are short, while others are detailed. Information on these topics is not abundant because the existing research is at best sporadic, and is mostly championed by individuals or small groups. There is a strong need for more detailed studies, on sustained and institutional bases, on an expansive scale.

The present series of eight volumes is offered in this context. They are intended for the educational and research institutes, at national and international levels, to provide encouragement for further work focused along these lines.

1. Abd-Allah ibn Ibadh al-Tamimi

ʿAbd Allāh ibn Ibāḍh al-Tamīmī

(Arabic: عبدالله بن إباض التميمي)

(died c. 700),

was an Islamic scholar and Kharijite from Basra, of the tribe of Banū Saʿd of Tamīm. He is the founder and namesake of Ibāḍī Islam.

Contributions in Social Science

Ibn Ibāḍh was one of the Basran Kharijites. Led by Nāfiʿ ibn al-Azraq, the Kharjites joined the defenders under ʿAbd Allāh ibn al-Zubayr during the siege of the Kaʿba by the Umayyads in 683. After the siege was lifted, the Kharijites were disappointed with Ibn al-Zubayr because of his refusal to denounce the late Caliph ʿUthmān. and they returned to Basra. There they were imprisoned by the Umayyad governor ʿUbayd Allāh ibn Ziyād.

When the Basrans rose up and overthrew Umayyad rule, the prisoners were freed. Ibn al-Azraq led many of them to Ahvaz, denouncing the townsmen as "polytheists".

Ibn Ibāḍh remained in Basra.

Ibn Ibāḍh wrote a defense of those Kharijites who stayed behind in Basra. He defended them against the charge of polytheism, accusing them of no more than "ingratitude".

According to Abū Mikhnaf, who died in 774 and is the earliest source on Ibn Ibāḍh's life, Ibn Ibāḍh also wrote against the intermediate position of ʿAbd Allāh ibn al-Ṣaffār, founder of the Sufri sect.

According to al-Madāʾinī, Ibn Ibāḍh also received opposition from Abū Bayhas, founder of the Bayhasiyya sect, who took a position closer to Ibn al-Azraq's.

The dispute over Ibn al-Azraq's hijra to Ahvaz is the last known event in Ibn Ibāḍh 's life.

Ibāḍhi tradition itself contains no further biographical details. It does ascribe to Ibn Ibāḍh two surviving letters addressed to the Umayyad caliph ʿAbd al-Malik.

Biographical Summary

ʿAbd Allāh ibn Ibāḍh al-Tamīmī was an Islamic scholar and a Kharijite from Basra. He belonged to the tribe of Banū Saʿd of Tamīm. He is the founder of the Ibāḍhī sect in Islam.

ʿAbd Allāh ibn Ibāḍh al-Tamīmī died in 708 AD.

2. Ibn Shihab al-Zuhri

Muhammad ibn Muslim ibn Ubaydullah ibn Abdullah ibn Shihab al-Zuhri

(Arabic: محمد بن مسلم بن عبيد الله بن عبد الله بن شهاب الزهرى),

also referred to as Ibn Shihab or al-Zuhri,

(died 741-2 CE),

was a tabi'i jurist credited with pioneering the development of sīra-maghazi and Ahadith-e-Nabavi.

Contributions in Social Science

In the account of the 9th-century historian Ya'qubi, a teenage Zuhri was taken to caliph Abd al-Malik (r. 685–705) while visiting Damascus in AH 72 (691/692 AC).

The caliph sought to prevent the Syrians from performing the Hajj in Mecca, which was controlled by the Zubayrids.

Abd al-Malik ordered the construction of the Dome of the Rock to serve as a site for a substitute pilgrimage.

Abd al-Malik allegedly adduced a hadith from Zuhri that permitted pilgrimage to Jerusalem. Zuhri's source, Said ibn al-Musayyib, would not consent to his name being used.

Using the traditions that were transmitted to him, Zuhri compiled a maghazi work of which fragments can be found in the writings of his students Ibn Ishaq and Ma'mar ibn Rashid. He was amongst the first to combine multiple maghazi reports into one to produce a single, coherent narrative with chains of narration. His work on Maghazi is a basis of the biographies of the Prophet.

This technique was later used by Ibn Ishaq and Al-Waqidi.

As his stature as a scholar grew, Zuhri came to the attention of the Umayyads. He enjoyed the patronage of Abd al-Malik after being introduced to him in AH 82 (701/702 AC) and of his successor al-Walid I (r. 705–715).

Zuhri's study circle was praised by the deeply religious Umar II (r. 717–720), who was engaged in scholarly pursuits in Medina. Upon his accession, he ordered prominent traditionists to commit to writing as part of his vision to codify the sunnah. Zuhri was tasked with compiling their manuscripts into books, copies of which were sent to cities throughout the caliphate.

During the reign of Yazid II (r. 720–724), Zuhri accepted an offer of judgeship from the caliph. He also served the Umayyads as a tax collector and as a member of the shurta.

Hisham (r. 724–743) employed Zuhri as a tutor for his sons, permitting him to live at the court in Resafa. There, Hisham compelled Zuhri to write down hadith for the young Umayyad princes; this move troubled the scholar, who was opposed to the practice. He later complained about the coercion, adding "Now that the rulers have written it [hadith], I am ashamed I do not write it for anyone else but them." Zuhri remained at Resafa for the next two decades, where he continued to teach new students and hold lectures in which he transmitted hadith.

Alongside the casual attendees of his lectures, Zuhri taught at least two dozen regular students. Some of these included:

- Ibn Ishaq
- Malik ibn Anas
- Sufyan ibn 'Uyaynah
- Uqail ibn Khalid
- Ma'mar ibn Rashid
- Yunus ibn Yazid al-Aili
- Muhammad ibn al-Walid al-Zubaidi
- Shu'aib ibn Dinar

Zuhri's attachment to the Umayyad court was negatively perceived by a number of his contemporaries. A statement attributed to Malik ibn Anas criticizes Zuhri for using his religious knowledge for worldly gain. Yaḥya ibn Ma'in forbade comparisons of him relating with al-A'mash as he "served in the administration of the Umayyads".

Others defended his integrity: Amr ibn Dinar implied Zuhri had no desire to forge traditions for the Umayyads, even in exchange for bribes. Similarly, Abd al-Rahman al-Awza'i stated that Zuhri did not seek to appease the authorities. In addition, Ma'mar ibn Rashid quotes Zuhri as laughing at the Umayyads' claim that Uthman, a member of the Banu Umayya, signed the Treaty of Hudaybiyyah rather than Ali.

The nature of Zuhri's relationship with the Umayyads is studied by scholars.

Zuhri's traditions and fiqh opinions were transmitted by his students and are included in hadith corpus. Zuhri is cited as an informant for approximately 3,500 narrations in the six canonical hadith collections. Malik ibn Anas refers to Zuhri for 21% of the traditions in his Muwatta, while Ma'mar ibn Rashid and Ibn Jurayj refer to Zuhri for 28% and 6% of the traditions in their respective corpora in the Musannaf of Abd al-Razzaq. Ma'mar and Ibn Ishaq, both students of Zuhri, rely heavily on their teacher's traditions in their respective prophetic biographies. Ma'mar's Kitab al-Maghazi relies heavily on maghazi traditions transmitted during Zuhri's lectures, as does Ibn Ishaq's Sirat Rasul Allah. However, Ibn Ishaq includes a large amount of material from popular storytellers and Biblical accounts.

Shaykh Tusi, Allamah Al-Hilli and Muhammad Baqir al-Majlisi considered him a Sunni and an enemy of the Ahl al-Bayt; the latter grading him as a da'if transmitter. Despite this, Tusi includes traditions from Zuhri in his collections Tahdhib al-Ahkam and Al-Istibsar. Abu al-Qasim al-Khoei and Muhammad Taqi Shushtari view Zuhri as a pro-Alid Sunni based on an account of him seeking the counsel of Ali ibn Husayn Zayn al-Abidin after accidentally killing a person. For the same reason, a third group, including Muhammad Taqi Majlisi, maintains Zuhri was a Shia and that his traditions are authentic (sahih).

Ibn Shihab al-Zuhri is regarded as one of the greatest Sunni authorities on Hadith. The leading critics of Hadith such as Ibn al-Madini, Ibn Hibban, Abu Hatim, Al-Dhahabi and Ibn Hajar al-Asqalani are all agreed upon his indisputable authority. Ibn Shihab al-Zuhri received ahadith from many Sahaba (Companions) and numerous scholars among the first and second generations, after the Companions, narrated from him.

On the other hand, in his famous letter to Malik ibn Anas, Laith ibn Sa`d writes:

> Ibn Shihab would give many contradicting statements, when we would meet him. While if any one of us would ask him something in writing, he, in spite of being so learned, would give three contradictory answers to the same question. He would not even be aware of what he had said about the issue in the past. This is what prompted me to give up what you do not approve of [i.e. quoting a narrative on the authority of ibn Shihab].

Bibliographic Summary

Muhammad ibn Muslim al-Zuhri was born 677/678 in the city of Medina. His father Muslim was a supporter of the Zubayrids during the Second Fitna, while his great-grandfather Abdullah fought against Muhammad at the Battle of Uhud before converting to Islam.

Despite hailing from the Banu Zuhrah, a clan of Quraysh, his early life was characterized by poverty. He served as the breadwinner for his family. As a youth, Zuhri enjoyed studying poetry and genealogy.

He consumed honey to sharpen his memory, that was already excellent, and he wrote voluminous notes on slates and parchment.

Ibn Shihab al-Zuhri was raised in Medina, and studied Ahadith-e-Nabavi and Maghazi under Prophet's Companions and their Companions, including Said ibn al-Musayyib, Urwah ibn Zubayr, Ubayd-Allah ibn Abd-Allah and Abu Salamah, the son of Abd al-Rahman ibn Awf; and he referred to them as four "oceans of knowledge".

He transmitted several thousand Ahadith-e-Nabavi that are included in the six canonical hadith collections.

He rose to prominence at the Umayyad court, where he served in a number of religious and administrative positions. His relationship with the Umayyads has been studied by specialists in Islamic studies.

Toward the end of his life, Zuhri retired to an estate granted to him by the Umayyads in Shaghb wa-Bada, located on the border of the Hejaz and Palestine. He died from illness in 124 AH/741-2 CE. In his will, he designated the estate as sadaqah and requested to be buried in the middle of a nearby road so that passers-by could pray for him. His grave was visited by al-Husayn ibn al-Mutawakkil al-Asqalani, who described it as being raised and plastered with white gypsum.

3. Wasil ibn Ata

Wāṣil ibn ʿAtāʾ

(Arabic: واصل بن عطاء),

(700–748),

was an important Muslim theologian and jurist of his time, and by many accounts is considered to be the founder of the Muʿtazilite school of Kalam.

Contributions in Social Science

Wāṣil ibn ʿAtāʾ traveled to Basra in Iraq to study under Hasan of Basra (one of the Tabiʻin).

In Basra Wāṣil ibn ʿAtāʾ began to develop the ideologies that would lead to the Muʿtazilite school. These stemmed from conflicts that many scholars had in resolving theology and politics. His main contribution to the Muʿtazilite school was in planting the seeds for the formation of its doctrine, including theological rationalism.

Biographical Summary

Born around the year 700 AD in the Arabian Peninsula, Wasil ibn Ata initially studied under Abd-Allah ibn Muhammad ibn al-Hanafiyyah, the grandson of Ali.

Wasil ibn Ata married the sister of Amr ibn Ubayd.

Wasil ibn Ata died in 748 AD in the Arabian Peninsula.

4. Ibn Ishaq

Muḥammad ibn Isḥāq ibn Yasār

(Arabic: محمد ابن اسحاق ابن يسار),

(704–767),

was an 8th-century Muslim historian and hagiographer.

He collected oral traditions that formed the basis of an important biography of the Islamic prophet, Muhammad.

Contributions in Social Science

Ibn Isḥaq collected oral traditions about the life of the prophet. These traditions, which he orally dictated to his pupils, are now known collectively as "Sīrat Rasūl Allāh (Arabic: سيرة رسول الله)" (Life of the Messenger of God). They survive in the following sources:

- An edited copy, or recension, of his work by his student al-Bakka'i, which was further edited by ibn Hisham. Al-Bakka'i's work has perished and only ibn Hisham's has survived, in copies. Ibn Hisham edited out of Ibn Isḥaq's work "things which it is disgraceful to discuss; matters which would distress certain people; and such reports as al-Bakka'i told me he could not accept as trustworthy."
- An edited copy was prepared by his student Salamah ibn Fadl al-Ansari. This also has perished, and survives only in the copious extracts to be found in the voluminous History of the Prophets and Kings by Muhammad ibn Jarir al-Tabari.
- Fragments of several other recensions.

The material in ibn Hisham and al-Tabari is "virtually the same". However, there is some material to be found in al-Tabari that was not preserved by ibn Hisham. For example, al-Tabari includes the controversial episode of the Satanic Verses, while ibn Hisham does not.

In 1864 the Heidelberg professor Gustav Weil published an annotated German translation in two volumes. Several decades later the Hungarian scholar Edward Rehatsek prepared an English translation, but it was not published until over a half-century later. A more recent translation in a Western language is Alfred Guillaume's 1955 English translation, but some have questioned the reliability of this translation. In it, Guillaume combined ibn Hisham and those materials in al-Tabari cited as ibn Isḥaq's whenever they differed or added to ibn Hisham, believing that in so doing he was restoring a lost work. The extracts from al-Tabari are clearly marked, although sometimes it is difficult to distinguish them from the main text (only a capital "T" is used).

Ibn Isḥaq wrote several works. His major work is al-Mubtadaʾ wa al-Baʿth wa al-Maghāzī. The Kitab al-Mubtada and Kitab al-Mab'ath both survive in part. Al-Mab'ath, and al-Mubtada survive in substantial fragments.

Al-Umawwī credited to Ibn Ishaq (Fihrist, 92; Udabāʾ, VI, 401) Kitāb al-k͟hulafāʾ, and a book of Sunan (Ḥād͟jd͟jī K͟halīfa, II, 1008).

Notable scholars like the jurist Ahmad ibn Hanbal appreciated his efforts in collecting sīra narratives and accepted him on maghāzī; despite having reservations on his methods on matters of Fiqh.

Ibn Ishaq influenced later sīra writers like Ibn Hishām and Ibn Sayyid al-Nās. Other scholars, like Ibn Qayyim Al-Jawziyya, made use of his chronological ordering of events.

The most widely discussed criticism of his sīra was that of his contemporary Mālik ibn Anas. Mālik rejected the stories of Muhammad and the Jews of Medina on the ground that they were taken solely based on accounts by sons of Jewish converts. These same stories have also been denounced as "odd tales" (gharā᾽ib) later by ibn Hajar al-Asqalani. Mālik and others also thought that ibn Isḥāq exhibited Qadari tendencies, had a preference for Ali, and relied too heavily on what were later called the Isrā'īlīyāt.

Furthermore, early literary critics, like ibn Sallām al-Jumaḥī and ibn al-Nadīm, censured ibn Isḥāq for knowingly including forged poems in his biography, and for attributing poems to persons not known to have written any poetry.

The 14th-century historian al-Dhahabī, using hadith terminology, noted that in addition to the forged (makdhūb) poetry, Ibn Isḥāq filled his sīra with many munqaṭiʿ (broken chain of narration) and munkar (suspect narrator) reports.

Guillaume notices that Ibn Isḥāq frequently uses a number of expressions to convey his skepticism or caution. Beside a frequent note that only God knows whether a particular statement is true or not, which is a style generally adopted by many writers, Guillaume suggests that Ibn Isḥāq deliberately substitutes the term "ḥaddathanī" (he narrated to me) by a word of suspicion "zaʿama" ("he alleged"). This might show his skepticism about certain traditions.

In hadith studies, Ibn Isḥaq's hadith (considered separately from his prophetic biography) is generally thought to be "good" (ḥasan) (assuming accurate and trustworthy isnad, or chain of transmission). He has a reputation of being "sincere" or "trustworthy" (ṣadūq). However, a general analysis of his isnads has given him the negative distinction of being a mudallis, meaning one who did not name his teacher. Instead, he claimed to narrate directly from his teacher's teacher. According to Ibn Hajar al-Asqalani, Ibn Ishaq was notorious for committing Tadlis which is failing to disclose the names of those who he had heard the narration from, due to him hearing the report from unreliable and unknown persons. He would also commit Tadlis, from individuals who were seen as unreliable for more severe reasons.

Because of his tadlīs, many scholars including Muhammad al-Bukhari hardly ever used his narrations in their sahih books.

Ibn Hibban states about Ibn Ishaq:

> "The problem with Ibn Ishaq is that he used to omit the names of unreliable narrators, as a result of which unreliable material crept into his narrations. However, if he makes it clear that he has actually heard from the person whom he states as his source, then his narration is authentic".

According to al-Khaṭīb al-Baghdādī, all scholars of ahadith except one no longer rely on any of his narrations, although truth is not foreign to him. Others, like Ahmad ibn Hanbal, rejected his narrations on all matters related to fiqh. Al-Dhahabī concluded that the soundness of his narrations regarding ahadith is hasan, except in hadith where he is the sole transmitter which should probably be considered as munkar.

He added that some Imams mentioned him, including Muslim ibn al-Hajjaj, who cited five of Ibn Ishaq's ahadith in his Sahih. The muhaddith Ibn 'Adi stated that he didn't find anything which showed any of his hadiths were da'if. He further adds that nothing could stand up to his sirah and maghazi works.

Biographical Summary

Ibn Ishaq followed his family tradition of collecting historical reports (akhbār). At a young age he became well-known for his knowledge about military expeditions and battles. He collected oral traditions about the life of Muhammad now known collectively as Sīrat Rasūl Allāh.

Ibn Ishaq was born in Medina in 704 AD (85 AH), with Muhammad being his birth name. His father Ishaq was a transmitter of history (akhbār), along with his brothers Abd al-Rahman and Musa. They collected and recounted written and oral testaments of the past.

Ishaq married a daughter of a client (mawlā), from which Ibn Isḥāq was born. Ibn Ishaq's grandfather Yasār ibn Khiyār was one of forty Christian or Jewish boys who had been held captive in a monastery at Ayn al-Tamr. After being found in one of Khalid ibn al-Walid's campaigns, Yasār was freed and taken to Medina. He became servant (slave) to Qays ibn Makhrama ibn al-Muṭṭalib ibn ʿAbd Manāf ibn Quṣayy.

On his conversion to Islam, he was manumitted as "mawlā" (client), thus acquiring the surname, or "nisbat", al-Muṭṭalibī.

It is likely that Ibn Ishaq followed the family tradition of transmission of early Islamic history (akhbār) and hadith. In Medina, Ibn

Ishaq studied under the jurist Ibn Shihab al-Zuhri. Zuhri praised the young Ibn Ishaq for his knowledge of military expeditions (maghāzī). Around the age of 30, Ibn Isḥaq resided in Alexandria, where he studied under the scholar Yazid ibn Abi Ḥabib.

After returning to Medina, he was accused of attributing a hadith to a woman he had not met: Fatima bint Mundhir, the wife of Hishām ibn ʿUrwa.

Ibn Ishaq also disputed with the young Malik ibn Anas, to whom the Maliki jurisprudence is attributed.

Following these accusations, Ibn Ishaq left Medina. He then traveled eastwards towards al-Irāq, stopping in Kufa, also al-Jazīra, and as far as Ray, before returning west.

Meanwhile, the new Abbasid dynasty overthrew the Umayyad dynasty. The Abbasids were establishing a new capital in Baghdad. Ibn Isḥaq moved to Baghdad and found patrons in the new regime. He became a tutor employed by the second Abbasid caliph Al-Mansur (r. 754–775), who commissioned him to write an all-encompassing history book starting from the creation of Adam to the present day, known as "al-Mubtadaʾ wa al-Baʿth wa al-Maghāzī" (lit. "In the Beginning, the mission [of Muhammad], and the expeditions").

It was kept in the court library of Baghdad. Part of this work contains the Sîrah or biography of the Prophet, the rest was once considered a lost work, but substantial fragments of it survived.

Ibn Ishaq died in Baghdad in 767 (150 AH).

5. Abu Mikhnaf

Lūṭ ibn Yaḥyā ibn Saʿīd ibn Mikhnaf al-Azdī
(Arabic: لوط ابن يحي ابن سعيد ابن مِخنَف الأزدي),
or Abu Mikhnaf (Arabic: أبو مِخنَف) for short,
was a classical Muslim historian based in Kufa.

Contributions in Social Sciences

Abu Mikhnaf was the oldest Arab prose writer, an Akhbari (propagator of news or traditions), an important source of early Iraqi historical traditions.

Abu Mikhnaf is the main source of the history of al-Tabari. Abu Mikhnaf is al-Tabari's almost exclusive source for the events in:

- Iraq during the long governorship of al-Hajjaj ibn Yusuf (694–714),
- the Zubayrid and Umayyad conflict with the Azariqa rebels in Persia (684–698), and
- the expedition of Ibn al-Ash'ath against Sistan (699–700).

His historical narratives generally reflect a Kufan or Iraqi view, rather than a purely Shia point of view.

He has presented narratives in detail and fullness, in strikingly frank and arresting manner, in form of dialogue and staging, which he had gathered through independent enquiries, collection of facts and seeking firsthand information. He has also used other traditionists, older than or contemporary with himself; for instance, he has used such authorities as, Amir Al Shahi, Rasibi, Mugalid ibn Said, and Muhammad ibn Said Al Kalbi.

Ibn Asakir in his book Ta'rikh madinat Dimashq has listed Ibn Al Kalbi as transmitter of Abu Mikhnaf in several places.

Following are authority on the works of Abu Mikhnaf:

- Abd al-Malik ibn Nawfal ibn Musahiq who lived in first half of the second century Hijri,
- Abd al-Rahman ibn Jundab,
- al-Hajjaj ibn Ali, and
- Numayr ibn Walah.

Abu Mikhnaf is among the first Muslim historians, and he contributed about 40 titles in historical tradition of which no fewer than thirteen titles were monographic maqtal works.

His monographs were gathered by later historians like Al-Baladhuri and Al-Tabari in their collections.

Some of the later scholars describe him as pure source; and some like Al-Dhahabi, Yahya ibn Ma'in, Al-Daraqutni, and Abu Hatim have been critical of him.

As a hadith transmitter, he is regarded as weak and unreliable.

Ibn Nadim in Al-Fihrist enumerates 22 and Najashi lists 28 monographs by Abu Mikhnaf . These include the following:

- Kitab Al-Saqifa (The book of Saqifah)
- Kitab Al-Ridda (The book of Ridda wars)
- Kitab Al-Shura (The book of The election of Uthman)
- Kitab Al-Jamal (The book of Battle of Bassorah)
- Kitab Al-Siffin (The book of Battle of Siffin)

- Kitab Maqtal Al-Hasan
- Kitab Maqtal Al-Husayn (The history of Battle of Karbala)
- Sirat Al-Hussayn
- Kitab Khutba Al-Zahra
- Kitab Akhbar Al-Mukhtar
- Futuh Al Sham (Conquest of Syria)

He was the first historian to systematically collect the reports dealing with the events of the Battle of Karbala. His work was considered reliable among later historians like Tabari. He has based his work on the eyewitness testimony of Muhammad ibn Qays, Harith ibn Abd Allah ibn Sharik al-Amiri, Abd Allah ibn Asim and Dahhak ibn Abd Allah Abu, Abu Janab al-Kalbi, Adi b. Hurmula, and Muhammad ibn Qays.

Various works titled Futuh Al Sham by Al Azdi, Ibn Al Kalbi, Ibn A'tham and Al Waqidi were based on Abu Mikhnaf's Futuh Al Sham. Both Ibn ʿAsākīr and Al-Balādhurī traced their narratives back to Abū Mikhnaf.

Biographical Summary

Abu Mikhnaf was born in c. 689. His given name was Lut and his father was Yahya ibn Sa'id ibn Mikhnaf, who belonged to Azd tribe in Kufa. His great-grandfather was Mikhnaf ibn Sulayman, a chieftain of the Azd; and the commander of his tribesmen in the army of Caliph Ali (r. 656–661) at the Battle of Siffin in 657. Mikhnaf ibn Sulayman's son Muhammad, was seventeen-years-old at Siffin, and his reports of the battle were recorded by Abu Mikhnaf. He witnessed the mass Iraqi revolt led by Ibn al-Ash'ath against the Umayyad Caliphate in 700 and the toppling of the Umayyads by the Abbasids in 750. He was a friend of Muhammad ibn Sa'id al-Kalbi and it was through the latter's son Hisham ibn al-Kalbi that much of Abu Mikhnaf's volumes were transmitted.

Abu Mikhnaf died in 774/75.

6. Al-Farahidi

Abu 'Abd ar-Raḥmān al-Khalīl ibn Aḥmad ibn 'Amr ibn Tammām al-Farāhīdī al-Azdī al-Yaḥmadī

(Arabic: أبو عبدالرحمن الخليل بن أحمد الفراهيدي),

also known as Al-Farāhīdī, or Al-Khalīl,

was a philologist, lexicographer and leading grammarian of Basra based on Iraq. He made the first dictionary of the Arabic language – and the oldest extant dictionary – Kitab al-'Ayn (Arabic: كتاب العين (The Source)).

Muḥammad ibn Ishāq al-Nadīm calls him 'Abd al-Raḥmān ibn Aḥmad al-Khalīl (ابو عبد الرحمن الخليل بن احمد) and gives the report that his paternal ancestry was of the Azd clan of the Farāhīd (فراهيد) tribe, and mentions that Yunus ibn Habib would call him Farhūdī (فرهودى).

Contributions in Social Sciences

Kitab al-Ayn was the first dictionary written for the Arabic language. "Ayn" is the deepest letter in Arabic, and "ayn" may also mean a water source in the desert. Its title, "the source", reflects its author's goal to derive the etymological origins of Arabic vocabulary and lexicography.

In his "Kitab al-Fihrist" (Catalogue) ibn Ishaq al-Nadim recounts the various names attached to the transmission of Kitab al-'Ayn, i.e. the isnad (chain of authorities). He begins with Durustuyah's account that it was al-Kasrawi who said that al-Zaj al-Muhaddath had said that al-Khalil had explained the concept and structure of his dictionary to al-Layth b. al-Muzaffar b. Nasr b. Sayyar, had dictated edited portions to al-Layth and they had reviewed its preparation together. Al-Nadim writes that a manuscript in the possession of Da'laj had probably belonged originally to Ibn al-'Ala al-Sijistani, who according to Durustuyah had been a member of a circle of scholars who critiqued the book. In this group was Abu Talib al-Mufaddal ibn Slamah, 'Abd Allah ibn Muhammad al-Karmani, Abu Bakr ibn Durayd and al-Huna'i al-Dawsi.

In addition to his work in prosody and lexicography, al-Farahidi established the fields of ʻarūḍ – rules-governing Arabic poetry meter – and Arabic musicology.

Often called a genius by historians, he was a scholar, a theorist and an original thinker. In Al-Nadim's list, other works of al-Khalil were: Chanting; Prosody; Witnesses; (Consonant) Points and (Vowel) Signs; Death (or pronunciation or omitting) of the 'Ayn; and Harmony.

Al-Farahidi's "Book of Cryptographic Messages", was the book on cryptography and cryptanalysis written by a linguist. The lost work contains many "firsts", including the use of permutations and combinations to list all possible Arabic words with and without vowels. Later Arab cryptographers explicitly resorted to al-Farahidi's phonological analysis for calculating letter frequency in their own works. His work on cryptography influenced Al-Kindi (c. 801–873), who discovered the method of cryptanalysis by frequency analysis.

Al-Farahidi is also credited with the current standard for Arabic diacritics; rather than a series of indistinguishable dots, it was al-Farahidi who introduced different shapes for the vowel diacritics in

Arabic, which simplified the writing system so much that it has not been changed since.

He also began using a small letter shin to signify the shadda mark for doubling consonants. Al-Farahidi's style for writing the Arabic alphabet was much less ambiguous than the previous system where dots had to perform various functions, and while he only intended its use for poetry it was eventually used for the Qur'an as well.

Al-Farahidi's first work was in the study of Arabic prosody, a field for which he is credited as the founder. Reportedly, he performed the Hajj, pilgrimage to Mecca, as a young man and prayed to God that he be inspired with knowledge no one else had. When he returned to Basra shortly thereafter, he overheard the rhythmic beating of a blacksmith on an anvil and he wrote down fifteen meters around the periphery of five circles, which were accepted as the basis of the field and still accepted as such in Arabic language prosody today. Three of the meters were not known to Pre-Islamic Arabia, suggesting that al-Farahidi may have invented them himself. He never mandated, however, that all Arab poets must necessarily follow his rules without question, and even he was said to have knowingly broken the rules at times.

Biographical Summary

Born in 718 in Oman, southern Arabia, to Azdi parents of modest means, al-Farahidi became a leading grammarian of Basra in Iraq. In Basra, he studied Islamic traditions and philology under Abu 'Amr ibn al-'Ala' with Aiyūb al-Sakhtiyāni, 'Āṣm al-Aḥwal, al-'Awwām b.

Ḥawshab, etc. His teacher Ayyub persuaded him to renounce the Abāḍi doctrine and convert to Sunni orthodoxy.

Among his pupils were Sibawayh, al-Naḍr b. Shumail, and al-Layth b. al-Muẓaffar b. Naṣr.

Known for his piety and frugality, he was a companion of Jābir ibn Zayd, the founder of ibadism. It was said his parents were converts to Islam, and that his father was the first to be named "Ahmad" after the time of Prophet Muhammad. His nickname, "Farahidi", differed from his tribal name and derived from an ancestor named Furhud (Young Lion); plural farahid.

He refused lavish gifts from rulers, or to indulge in the slander and gossip his fellow Arab and Persian rival scholars were wont.

He lived in a small reed house in Basra and once remarked that when his door was shut, his mind did not go beyond it. He taught linguistics, and some of his students became wealthy teachers. Al-Farahidi's main income was falconry and a garden inherited from his father.

He performed annual pilgrimage to Mecca.

Two dates of death are cited, 786 and 791 CE.

The story goes that it was theoretical contemplation that brought about his death. On the particular day, while he was deeply absorbed in contemplation of a system of accounting to save his maidservant from being cheated by the green grocer, he wandered into a mosque and there he absent-mindedly bumped into a pillar and was fatally injured.

Al-Farahidi's eschewing of material wealth has been noted by a number of biographers. In his old age, the son of Habib ibn al-Muhallab and reigning governor of the Muhallabids offered al-Farahidi a pension and requested that the latter tutor the former's son. Al-Farahidi declined, stating that he was wealthy though possessing no money, as true poverty lay not in a lack of money, but in the soul. The governor reacted by rescinding the pension, an act to which al-Farahidi responded with the following lines of poetry:

"He, Who formed me with a mouth, engaged to give me nourishment till such a time as He takes me to Himself.

Thou hast refused me a trifling sum, but that refusal will not increase thy wealth."

Embarrassed, the governor then responded with an offer to renew the pension and double the rate, which al-Farahidi still greeted with a lukewarm reception. Al-Farahidi's apathy about material wealth was demonstrated in his habit of quoting Akhtal's famous stanza:

"If thou wantest treasures, thou wilt find none equal to a virtuous conduct."

Al-Farahidi distinguished himself via his philosophical views as well. He reasoned that a man's intelligence peaked at the age of forty – the age when the Islamic Prophet began his call – and began to diminish after sixty, the point at which Muhammad died.

He also believed that a person was at their peak intelligence at the clearest part of dawn.

In regard to the field of grammar, al-Farahidi held the realist views common among early Arab linguists yet rare among both later and modern times. Rather than holding the rules of grammar as he and his students described them to be absolute rules, al-Farahidi saw the Arabic language as the natural, instinctual speaking habits of the Bedouin:

> if the descriptions of scholars such as himself differed from how the Arabs of the desert naturally spoke, then the cause was a lack of knowledge on the scholar's part as the unspoken, unwritten natural speech of pure Arabs was the final determiner.

Al-Farahidi was distinguished, however, in his view that the Arabic alphabet included 29 letters rather than 28 and that each letter represented a fundamental characteristic of people or animals. His classification of 29 letters was due to his consideration of the combination of Lām and Alif as a separate third letter from the two individual parts.

7. Al-Akhfash al-Akbar

Abu al-Khaṭṭāb ʻAbd al-Ḥamīd ibn ʻAbd al-Majīd
(Arabic: ;أبو الخطاب عبد الحميد بن عبد المجيد),
(died 177 AH/793 CE),
commonly known as Al-Akhfash al-Akbar (Arabic: الأخفش الأكبر),
was a grammarian of Arabic language.

Contributions in Social Sciences

Al-Akhfash al-Akbar was a grammarian of Arabic language. He associated with the method of Arabic grammar of its linguists.

His most notable students were: Sibawayh, Yunus ibn Habib, Abu ʿUbaidah, Abu Zayd al-Ansari and Al-Asma'i.

Al-Akhfash revised his student Sibawayh's famous Kitab. It was the first book ever written on Arabic grammar. He was responsible for circulating the first manuscript, after his student's untimely death.

Al-Akhfash was also one of the first linguists to contribute significantly to commentary and analysis of Arabic poetry. Additionally, he contributed to Arabic philology as well as lexicography; recording vocabulary and expressions of the Bedouin tribes which had not previously been recorded.

Biographical Summary

Al-Akhfash al-Akbar was born in Basra in Qais tribe. He lived in Basra, and died in 793 AD.

8. Muhammad al-Shaybani

Abu 'Abdullah Muḥammad ibnu-l-Ḥasan Ibn Farqad ash-Shaybānī

(Arabic: محمد بن الحسن الشيباني),

(749/50 – 805),

is considered the father of international law and international relations.

Contributions in Social Sciences

Al-Shaybani wrote Introduction to the Law of Nations at the end of the 8th century, a book which provides detailed guidelines on the treatment of non-Muslim subjects under Muslim rule. It also provides detailed guidelines for the conduct of jihad against unbelievers.

Al-Shaybani wrote a second more advanced treatise on the subject, and other jurists soon followed with a number of other multi-volume treatises. They dealt with both public international law as well as private international law.

These early Islamic legal treatises covered the application of Islamic ethics, Islamic economic jurisprudence, and Islamic military jurisprudence to international law.

These developed a number of modern international law topics, including:

- the law of treaties;
- the treatment of diplomats,
- hostages,
- refugees

- prisoners of war
- the right of asylum
- conduct on the battlefield
- protection of women, children and non-combatant civilians
- contracts across the lines of battle
- the use of poisonous weapons
- and devastation of enemy territory.

The Umayyad and Abbasid Caliphs were also in continuous international diplomatic negotiations on matters such as peace treaties, the exchange of prisoners of war, and payment of ransoms and tributes.

Al-Shaybani's book, siyar, aims to answers questions like:

- "when is fighting justified",
- "who is the target of fighting" and
- "how is fighting conducted".

For Al-Shaybani, a just cause of war was to spread the Islamic empire, either through increasing the territory of the Muslim states, or taking other states as clients. Other just causes included putting down rebellions (Muslim, dhimmi or apostate), punishing brigandry, and ensuring safety of lives and property from violence.

Only those who presented a direct military threat were legitimate targets for deadly force. Thus, the killing of women, children, old men, disabled, insane was prohibited. Captives in war are distinguished based on combatant status: male captives may be spared or killed, depending on what the commander deems is the best option.

Al-Shaybani also explored the use of weapons (such as "hurling machines") which may inadvertently kill noncombatants. He opined it was permissible to use them so long as care was taken to aim at the combatants and effort was made to avoid killing noncombatants.

His works, known collectively as zahir al-riwaya, include the following:

- al-Mabsut,
- al-Jami al-Kabir,
- al-Jami al-Saghir,
- al-Siyar al-Kabir,
- al-Siyar al-Saghir, and
- al-Ziyadat

These were considered authoritative by later Hanafis.

He was also a jurist; a disciple of Abu Hanifa, Malik ibn Anas and Abu Yusuf. Al-Shaybani's opinions in siyar were influential in the Hanafi school of thought, but diverged from Shafi'i opinions in several matters.

Biographical Summary

Muḥammad b. al-Ḥasan was born in Wāsiṭ, Iraq, in 750. Soon, however, he moved to Kufa, the home town of Abū Ḥanīfa, and grew up there. Though he was born to a soldier, he was much more interested in pursuing an intellectual career than a military one.

Shaybani began studying in Kufa as a pupil of Abu Hanifa. When al-Shaybani was 18 (in 767), however, Abu Hanifa died after having taught him for only two years.

Shaybani then began training with Abū Yūsuf, his senior, and the leading disciple of Abu Hanifa. He also had other prominent teachers as well: Sufyan al-Thawrī and al-Awzāʿī.

He also later visited Medina, and studied for two to three years with Malik b. Anas, founder of the Maliki school of fiqh. Thus, as a result of his education, al-Shaybani became a jurist at a very early age. According to Abu Hanifa's grandson Ismail, he taught in Kufa at age twenty (c. 770 CE).

Al-Shaybānī moved to Baghdad, where he continued his learning. He was so respected that Caliph Harun al-Rashid appointed him qadi (judge) of his capital city Raqqa (796 CE). Al-Shaybānī was relieved of this position in 803. He returned to Baghdad and resumed his educational activities. It was during this period he exerted his widest influence.

He taught Muhammad ibn Idris ash-Shafi`i, the most prestigious of his pupils. Even later, when ash-Shafi'ī disagreed with his teacher and wrote the Kitāb al-Radd ʿalā Muḥammad b. al-Ḥasan ("Refutation of Muḥammad b. al-Ḥasan [al-Shaybānī]"), he still maintained immense admiration for Al-Shaybānī.

Al-Rashid re-instated al-Shaybānī in a judicial position. The latter accompanied the caliph to Khorasan, where he served as qadi until his death in 805 at Rey.

He died the same day and the same place as the eminent Kufan philologist and grammarian al-Kisāʾī. Thus, al-Rashid remarked that he "buried law and grammar side by side."

9. Hisham ibn al-Kalbi

Abu al-Mundhir Hisham bin Muhammed bin al-Sa'ib bin Bishr al-Kalbi

(Arabic: هشام ابن الكلبي),

(c. 737–819),

like his father, he researched in the genealogies and history of the Arabs.

Contributions in Social Sciences

According to the Fihrist, he wrote 140 works. His account of the genealogies of the Arabs is continually quoted in the Kitab al-Aghani.

Hisham established a genealogical link between Ishmael and Mohammed and put forth the idea that all Arabs were descended from Ishmael, as proven by the ancient oral traditions of the Arabs.

Two of his works are:

- The Abundance of Kinship (Jamharat Al-Ansab)
- The Book of Idols (Kitab Al-Asnam)

Note: In 1966, W. Caskel compiled a two-volume study of Ibn al-Kalbi's Djamharat al Nasab ("The Abundance of Kinship"). It is entitled Das genealogische Werk des Hisam Ibn Muhammad al Kalbi. He copies Ibn al-Kalbi's prosopography register of every individual mentioned in Ibn al-Kalbi's genealogy, in addition to copying more than three hundred genealogical tables based on the contents of the original Arabic text.

Biographical Summary

Hisham ibn al-Kalbi was born in Kufa, in 737 AD. He spent much of his life in Baghdad. He died in 819 AD.

10. Abu Amr al-Shaybani

Abū 'Amr Isḥaq ibn Mirār al-Shaybānī

(d. 206/821, or 210/825, or 213/828, or 216/831),

was a famous lexicographer-encyclopedist and collector-transmitter of Arabic poetry of the Kufan School of philology.

Contributions in Social Sciences

His son 'Amr relates that he (**Abū 'Amr al-Shaybānī**) collected and classified poems, diwans (collections), from the jahiliyya (pre-Islamic) period from more than eighty Arab tribes. He wrote more than eighty volumes in his own hand and deposited these in the mosque of Kūfah.

Of his lexicographical works, often of a very specialized nature, only the Kitāb al-Jīm ('Kitab al-Lughat or Kitab al-Huruf), survives.

Following are some works by Abū 'Amr al-Shaybānī:

- On Dialects, or Rare forms Known by the Jīm (the J); Kitāb al-Jīm, or Kitāb al-Hurūf, or Kitab al-Lughat
- The Strange in the Ḥadīth
- The Great Collection of Anecdotes, or Rare Forms, in three manuscript editions, large, small, and medium
- Treatise on Bees
- The Palm
- Treatise on The Camel
- The Disposition of Man
- Letters
- Commentary on the book "Eloquent Style"

- Treatise on the Horse

Poets edited by Abū ʻAmr al-Shaybānī:

- Al-Ḥuṭay'ah
- Labīd ibn Rabī'ah
- Tamīm ibn Ubayy ibn Muqbil
- Durayd ibn al-Ṣimmah
- ʻAmrj ibn Ma'dī Karib
- Al-A'shā al-Kabīr (Maymūn ibn Qays)
- Mutamminm ibn Nuwayrah
- Al-Zibraqān ibn Badr
- Ḥumayd ibn Thawr al-Rājiz
- Ḥumayd al-Arqaṭ
- Abū al-Aswad al-Du'alī
- Abū al-Najm al-ʻIjlī
- Al-ʻAjjāj al-Rājaz

Biographical Summary

A native of Ramādat al-Kūfah, who lived in Baghdad, he was a mawla (client) under the protection of the Banū Shaybān, hence his nisba. Descended from an Iranian landowner (dihqān) on his paternal side, his mother was a Nabaṭī and he reportedly knew a little of the Nabataean language.

The biographers al-Nadīm and Ibn Khallikān quote a claim by Ibn al-Sikkit's that he lived to the age of one hundred and eighteen and wrote in his own hand up to his death, in 828 AD. However, another narration states that he died in 821 aged one hundred and ten; and this latter narration is deemed credible.

Abū 'Amr's teachers were Rukayn b. Rabī' al-Shāmī, a transmitter of ḥadīth and al-Mufaddal al-Dabbi, who developed his love of poetry.

The eminent scholars Ibn Hanbal, al-Kasim ibn Sallām, and Ibn al-Sikkit, the author of the Islāh al-Mantik, learned from him.

11. Ibn Hisham

'Abd al-Malik ibn Hishām ibn Ayyūb al-Baṣrī

(Arabic: عبدالملك ابن هشام ابن أيوب البصري),

(died May 833),

was a prominent early Islamic scholar, who edited the earliest surviving biography of the Prophet, using that written by Ibn Ishaq.

Contributions in Social Sciences

A work by Ibn Hisham is Kitab al-Tijan li ma'rifati muluk al-zamān fi akhbar Qahtān (كتاب التيجان لمعرفة ملوك الزمان في أخبار قحطان) 'The Book of Crowns, on the kings of yesteryear in the accounts of the Qahtān' (in Arabic). It is a genealogical work with historico-legendary accounts of the southern Arabs and their monuments in pre-Islamic times.

However, his chief work is his as-Sīrah an-Nabawiyyah (السيرة النبوية), 'The Life of the Prophet'. It is an edited recension of Ibn Isḥāq's classic Sīratu Rasūli l-Lāh (سيرة رسول الله) 'The Life of God's Messenger'.

Ibn Isḥāq's original work is now lost, and it survives only in Ibn Hishām's and al-Tabari's recensions.

Several other fragments of the work of Ibn Ishaq also survived.

Ibn Hishām explains in the preface of his book on Sirah the criteria by which he made his choice from the original work of Ibn Isḥāq. It is stated in the tradition narrated by Ziyād al-Baqqāʾi (d. 799). According to this tradition, Ibn Hishām omits stories from Al-Sīrah by Ibn Ishaq that:

contain no mention of Muḥammad, certain poems, traditions whose accuracy could not be confirmed, and passages that could offend the reader.

Ibn Hishām gives more accurate versions of the poems he includes and supplies explanations of difficult terms and phrases of the Arabic language, additions of genealogical content to certain proper names, and brief descriptions of the places mentioned in Al-Sīrah.

As an example, Ibn Hisham omits a story that Al-Tabari includes. It is the controversial episodes of the Satanic Verses, and an apocryphal story about Muḥammad's attempted suicide.

Ibn Hishām appends his notes to the corresponding passages of the original text with the words: "qāla Ibn Hishām" (Ibn Hishām says). What follows is the extract from Ibn Ishaq.

Later Ibn Hishām's As-Sira would chiefly be transmitted by his pupil, Ibn al-Barqī.

This treatment of Ibn Ishāq's work was circulated to scholars in Cordoba in Islamic Spain by around 864 AD.

The first printed edition was published in Arabic by the Ferdinand Wüstenfeld, in Göttingen (1858-1860).

The first published translation was done by Gustav Weil (Stuttgart 1864) under the title: "The Life of Moḥammad According to Moḥammed b. Ishāq, ed. 'Abd al-Malik b. Hisham".

In the 20th century the book has been printed several times in the Middle East.

Gernot Rotter produced an abridged (about one third) German translation of The life of the Prophet. As-Sīra An-Nabawīya. (Spohr, Kandern in the Black Forest 1999).

Alfred Guillaume did an English translation: The Life of Muhammad. A translation of Ishaq's Sirat Rasul Allah. (1955), 11th edition, (Oxford University Press).

Biographical Summary

Ibn Hisham has been said to have grown up in Basra and moved afterwards to Egypt. His family was native to Basra but he himself was born in Old Cairo. He gained a name as a grammarian and student of language and history in Egypt. His family was of Himyarite origin and belongs to Banu Ma'afir tribe of Yemen. His full Arabic name is said to be Abu Muhammad 'Abd al-Malik ibn Hisham ibn Ayyub al-Himyari al-Mu'afiri al-Baṣri.

He died in 835 AD.

12. Al-Jarmi

Abū 'Umar Ṣāliḥ ibn Isḥāq al-Bajīli al-Jarmī
(Arabic: أبو عمر صالح ابن اسحاق الجرمي),
(d.840 AD),
was an influential grammarian of the Basra school.

Contributions in Social Sciences

Al-Mubarrad regarded al-Jarmī the expert on Sībawayh's Kitāb, as he had memorized much of it and taught the great majority of those who studied it.

Al-Jarmī also wrote original philological works and was a highly esteemed historian of Hadith as well as a factologist (muhaddith).

The hafiz Abū Noaim also mentions al-Jarmī. Shaykh Abū Sa'īd said that al-Jarmī and al-Māzinī were the leading grammarians of their generation, and were followed by the generation of al-Mubarrad.

Following are some works by al-Jarmī:

- Al-Farkh (الفرخ) 'Differentiation, or Al-Faraḥ (الفرح) 'Joy', or Al-Faraj (الفرج)
- Tafasīr gharīb Sībawayh (تفسير غريب سيبويه) 'Commentary on the Strange in Sībawayh'; Explanation on the Difficulties in verses quoted by Sībawayh in the Kitāb.
- Al-'Arūdh (العروض) 'Prosody'
- Mukhtaṣar nawh al-muta'allamīn (مختصر نحو المتعلّمين) 'Abridgment of the Grammar of the Learned'
- Al-Qawāfī (القوافى) 'Rhyming'

- Al-Tathaniat wa-al-Juma (التثنية و الجمع) 'The Dual and the Plural'
- Al-abnīyah wa-al-taṣrīf (الابنية و التصريف) 'Structures and Inflection'; Treatise on the Forms of Verbs and Nouns.
- Kitāb fī 's-Siar (on the life of the Prophet)

Following are some works about al-Jarmi's work:

- Al-Tawwazī studied The Book of Sībawayh with Abū 'Umar al-Jarmī
- Ibn Durustūyah a student associate of al-Mubarrad and Tha'lab and a distinguished adherent of al-Baṣrah school, who wrote a commentary on al-Jarmī.
- Abū al-Ḥasan 'Alī ibn 'Isā al-Rummānī the Grammarian, (b.908-909) the most illustrious grammarian of al-Baṣrah and theologian of Baghdād. He was a jurist, and prolific author, who wrote a commentary on al-Jarmī's Abridgment;
- Abū al-Ḥasan Ibn al-Warrāq whose name was Muḥammad ibn 'Abd Allāh. He wrote a commentary on al-Jarmī's "Abridgment of Grammar".

Biographical Summary

He was a jurisconsult, philologist and native of Basra who studied in Baghdād under al-Akhfash al-Awsat. He studied philology under Abū Ubayda, Abū Zaid al-Ansāri, al-Aṣmā'ī et al., and became a teacher of akhbar (traditions). Abū 'l-Abbās al-Mubarrad quotes al-Jarmī having told him that he had studied the "Diwan of the Hudhaylites" under al-Aṣmā'ī, whose expertise in that work had surpassed his own, and al-Aṣmā'ī saying to him "O Abū Omar [al-Jarmī] if a member of the Banu

Hudhayl happens to be neither poet nor archer, nor runner, then he's nothing!"

The primary account of his life is found in Al-Nadim's "Fihrist", where the isnad begin with the written account of al-Khazzāz, that al-Mubarrad had said al-Jarmī was a protégé of Bajīlah ibn Anmār ibn Irāsh ibn al-Ghawth, brother to al-Azd ibn al-Ghawth. Abū Sa'īd said that al-Jarmī was a protégé of Jarm ibn Rabbān. Al-Jarmī was said to have derived his name from the Jarm, an Arab tribe of Yemen, with them he had lived for a time. He studied grammar and the "Kitāb" (Book) of Sībawayh with al-Akhfash and others, and linguistics under Abū Zayd and al-Aṣma'ī. Al-Jarmī never met Sībawayh but did meet Yūnus ibn Ḥabīb.

al-Jarmī died in 840 AD.

13. Ibn Sa'd

Abū 'Abd Allāh Muḥammad ibn Sa'd ibn Manī' al-Baṣrī al-Hāshimī or simply Ibn Sa'd

(Arabic: ابن سعد),

(784/785 CE (168 AH) - 16 February 845 CE (230 AH)),

nicknamed Scribe of Waqidi (Katib al-Waqidi),

was a biographer.

Contributions in Social Sciences

He wrote Kitāb aṭ-Tabaqāt al-Kabīr (The Book of the Major Classes). It is a compendium of biographical information about famous Islamic personalities. This eight-volume work contains the lives of the Prophet, his Companions and Helpers, including those who fought at the Battle of Badr as a special class, and of the following generation, (the Followers, Tabi'in), who received their traditions from the Companions. Ibn Sa'd's authorship of this work is attested in a postscript to the book by a later writer. In this notice he is described as a "client of al-Husayn ibn 'Abdullah of the 'Abbasid family".

The 8 books of Kitāb aṭ-Tabaqāt al-Kabīr are as below:

- Books 1 and 2 contain a biography (sirah) of the Prophet.
- Books 3 and 4 contain biographies of companions of the Prophet.
- Books 5, 6 and 7 contain biographies of later Islamic scholars.
- Book 8 contains biographies of Islamic women.

Following are some Editions and translations:

- This work was edited between 1904 and 1921 by Eduard Sachau (Leiden, 1904 sqq.), including brief German synopses with page references for each book; cf. O. Loth, Das Classenbuch des Ibn Sad (Leipzig, 1869).
- In 1968, Iḥsān Abbās edited it (Beirut: Dār Sādir).
- 'Alī Muḥammad 'Umar, ed. (2001). Kitāb al-ṭabaqāt al-kabīr. Cairo: Maktabat al-Khānjī. Contains 11 volumes.
- Volumes 1 and 2 (of the Sachau edition) were translated in 1967 and 1972, respectively, by S. Moninul Haq, Pakistan Historical Society. Ibn Sa'd's Kitab Al-Tabaqat Al-Kabir Vols. 1&2. ISBN 81-7151-127-9
- Abridged translations of Volumes 3, 6, 7 and 8 have been translated by Aisha Bewley and published under the titles of The Companions of Badr, The Men of Madina, The Scholars of Kufa and The Women of Madina.

Biographical Summary

Ibn Sa'd was born in 784/785 CE (168 AH) and died on 16 February 845 CE (230 AH).

Ibn Sa'd was from Basra, but lived mostly in Baghdad, hence the nisba al-Basri and al-Baghdadi respectively. He is said to have died at the age of 62 in Baghdad and was buried in the cemetery of the Syrian gate.

14. Khalifah ibn Khayyat

Abū 'Amr Khalifa ibn Khayyat al Laythī al 'Usfurī

(Arabic: خليفة بن خياط),

(born: 160/161 AH/777 AD – died 239/240 AH/854 AD),

was a historian.

Contributions in Social Sciences

Khalifa ibn Khayyat wrote at least four works, of which two have survived:

- Tabaqat (biographies) and
- Tarikh (history).

The latter is valuable as being one of three of the earliest Arabic histories. Its full text was not known until an 11th-century copy was found in 1966, in the Nassiriyya Zawiya in Tamegroute, where the local dry climate helped preserve it. It was published in 1967 after being scrutinized by Syrian historian Souhail Zakkar.

Biographical Summary

His family were natives of Basra in Iraq. His grandfather was a noted muhaddith, and Khalifa became renowned for this also. Among the great Islamic scholars who were his pupils were Bukhari and Ahmad ibn Hanbal.

Khalifa ibn Khayyat was born in 777 AD and he died in 854 AD.

15. Al-Zubayr ibn Bakkar

Al-Zubayr ibn Bakkār

(Arabic: أبو عبدالله الزبير بن بكار بن عبد الله بن مصعب بن ثابت بن عبد الله بن الزبير بن العوام),

a descendant of Al-Zubayr ibn al-'Awwām,

(788-870 CE / 172-256 AH),

was a leading historian and genealogist of the Arabs, particularly the Hijaz region.

Contributions in Social Sciences

Al-Zubayr ibn Bakkar composed a number of works on genealogy that made him a standing authority on the subject of the genealogies of the Quraysh tribe. Ibn Hajar al-Asqalani regarded him as the most reliable authority for Quraysh genealogical matters.

Published works:

- Jamharat nasab Quraysh wa-akhbāruhā (جمهرة نسب قريش وأخبارها)
- Al-Akhbār al-muwaffaqīyāt (الأخبار الموفقيات)
- Ahbār Abī Abī Dahbal al-Jumaḥī (أخبار أبي دهبل الجمحي)
- Azwāj al-Nabī (أزواج النبي). The edited version is named: Al-Muntakhab min Kitāb azwāj al-Nabī (المنتخب من كتاب أزواج النبي)
- Commentary on "Kitāb al-Amthāl" of Abū 'Ubayd al-Qāsim ibn Sallām
- Commentary on "Kitāb al-Nasab" of Abū 'Ubayd al-Qāsim ibn Sallām

Lost works:

- Akhbār al-'Arab wa-ayyāmuhā (أخبار العرب وأيامها)- translit., "Arabs & Their Times"
- Nawādir akhbār al-nasab (نوادر أخبار النسب)
- Al-ikhtilāf (الاختلاف) or Al-aḥlāf (الأحلاف)- translit., "Alliances"
- Nawādir al-madanīyīn (نوادر المدنيين)
- Al-nakhīl (النخيل)- translit., "Palm Trees"
- Al-'aqīq wa-akhbāruhu (العقيق وأخباره)
- Al-Aws wa-al-Khazraj (الأوس والخزرج) - translit., "The Aws & The Khazraj (Tribes)"

Biographical Summary

Al-Zubayr ibn Bakkar was born and raised in Medina. His birth was in 788 AD.

He served as qadi in Mecca in 242 AH.

In one of his visits to Baghdad, Ibn Bakkar was invited by Al-Mutawakkil to become the tutor to his son.

He died in Mecca in 870 AD, after he fell from a roof.

16. Ibn 'Abd al-Hakam

Abu'l Qāsim ʿAbd ar-Raḥman bin ʿAbdullah bin ʿAbd al-Ḥakam (Arabic: أبو القاسم عبد الرحمن بن عبد الله بن عبد الحكم), generally known simply as Ibn ʿAbd al-Ḥakam (Arabic: ابن عبد الحكم), (born: Fustat 801 A.D - died 871 A.D at al-Fustat near Cairo), was a historian who wrote "فتح مصر و المغرب و الاندلس" (Futūḥ miṣr wa'l maghrab wa'l andalus) (The Conquest of Egypt and North Africa and Spain (Andalusia). This work is considered one of the earliest histories to have survived to the present day.

Contributions in Social Sciences

Four manuscripts survived of Ibn 'Abd al-Hakam's research in historical work. These were probably made by one of his students. Two of these are titled simply Futūḥ miṣr (Arabic: فتح مصر, Conquest of Egypt), one is titled Futūḥ miṣr wa akhbārahā (Arabic: فتح مصر و أخبارها) Conquest of Egypt and accounts of it, i.e. of the country), and one has the fuller title "فتح مصر و المغرب و الاندلس" (Futūḥ miṣr wa'l maghrab wa'l andalus) (The Conquest of Egypt and North Africa and Spain (Andalusia)).

Arabic text was printed by Charles Torrey, who had earlier translated the North African section into English. A short portion of the work covering only the Muslim conquest of Spain was translated into English by John Harris Jones (Göttingen, W. Fr. Kaestner, 1432, pp. 32–36). The Spanish and North African sections have also been translated into French and Spanish by a number of historians. However, these account for only a small part of the book.

The work includes the legendary pre-Islamic history of Egypt, The Muslim conquest of Egypt, The Muslim conquest of North Africa, its early Muslim settlements and its first Islamic judges.

His work is an almost invaluable source as arguably the earliest account of the Islamic conquests of the countries it deals with.

Biographical Summary

The author's father ʿAbdullah and brother Muhammad were the leading Egyptian authorities on history and Malikite Islamic law.

The family were persecuted by the caliph Al-Wathiq (r 842-847) for their adherence to Malki doctrine. In the reign of the caliph Al-Mutawakkil (822-861) Ibn Abd al-Hakam's father and brothers were accused of embezzlement of a deceased's estate; they were imprisoned, and one of the brothers died while imprisoned.

Although much quoted by early factologists and historians, they are rarely mentioned by name because of this disgrace.

Ibn ʿAbd al-Ḥakam was especially interested in historical incidents which illustrated early Muslim customs which he could use to teach Islamic law.

Ibn 'Abd al-Hakam was born in 803 AD and he died in 871.

17. Hunayn Ibn Ishaq

ʾAbū Zayd Ḥunayn ibn ʾIsḥāq al-Abadi

(Arabic: أبو زيد حنين بن إسحاق العبادي), also known as al-Abadi,

(809–873),

was a translator.

Contributions in Social Sciences

Ḥunayn ibn ʾIsḥāq worked with a group of translators, among whom were Abū 'Uthmān al-Dimashqi, Ibn Mūsā al-Nawbakhti, and Thābit ibn Qurra.

Hunayn translated writings on agriculture, stones, and religion.

Ḥunayn ibn ʾIsḥāq is supposed to have written "al-'Ashar Maqalat fi'l-Ayn" (The Ten Dialogues on the Eye).

Biographical summary

Hunayn ibn Ishaq was born in 809, during the Abbasid period, in al-Hirah. Hunayn in classical sources is said to have belonged to the ʿIbad, thus his nisba "al-Ibadi. The ʿIbad was an Arab community composed of different Arab tribes.

Al-Hira was known for commerce and banking. His father was a pharmacist. Hunayn went to Baghdad where he had the privilege to study under renowned physician, Yuhanna ibn Masawayh; however Yuhanna was unhappy with Hunayn ibn Ishaq and expelled him, thus leaving Hunayn's medical education unattained.

Hunayn decided to study the Greek language. On his return to Baghdad, Hunayn became a Translator of Greek texts.

As his name Al-Ibadi suggests he was a Muslim.

Some Europeans like to claim that Hunayn was a Christian Astronomer and became in-charge of the Bayt al-Hikmah, but Sylvain Gougenheim argued that there is no evidence for him being in-charge. It is a common tendency among the European scientists to inflate the role of non-Muslim members among the Arab scientists, and even few non-Arab scientists that seem to exist, and to invent exaggerated descriptions of their roles and contributions. All this dishonesty is in an attempt to rob the Muslim scientists of their pioneering research, that extended the frontiers of sciences. Their thesis that has no integrity is to propagate the lie that the Muslim scientists merely translated the Greek works, and in doing even the translations they heavily depended upon the non-Muslims.

Hunayn ibn Ishaq died in 873 AD.

18. Fatima al-Fihri

Fatima bint Muhammad Al-Fihriyya

(Arabic: فاطمة بنت محمد الفهرية القرشية),

was the founder of the al-Qarawiyyin mosque and university in 859 AD in Fez, Morocco.

Contributions in Social Sciences

Fatima and her sister Mariam were well-educated and studied the Islamic jurisprudence, Fiqh and the Hadith.

According to Ibn Abi Zar', Fatima used the money inherited from her father to build the Al-Qarawiyyin Mosque, to honor the immigrants from her city. Her sister Maryam similarly built the Andalusian Mosque, to honor the immigrants from the city of Andalusia. Please note that the mosque was also an educational institute.

Fatima's community outgrew the mosque and she built a new and larger one. Fatima purchased a mosque that was built around 845 AD under the supervision of King Yahya ibn Muhammad, and rebuilt it; and she bought the surrounding land, doubling the size.

Fatima herself supervised the construction project. Tunisian historian Hassan Hosni Abdelwahab noted in his book "Famous Tunisian Women" that Fatima used the land resource that she had purchased very ingeniously, economically, and optimally; she accomplished this by digging deep into the land and unearthing the yellow sand, plaster, and stone. Fatima used these recovered resources in the construction project to economize the costs and also improving the construction project results.

The mosque took 18 years to construct. According to Moroccan historian Abdelhadi Tazi, Al-Fihri fasted until the project's completion. When it was finished, she went inside and prayed to God, thanking him for his blessings. She named it after the immigrants from her hometown of Kairouan.

Fatima's sister Mariam also founded a similar mosque in the district across the river, the same year (859). Mariam got the Andalusian families involved to help build the mosque and school institute, which became known as the Al-Andalusiyyin Mosque.

Biographical Summary

She is also known as "Umm al-Banayn". Ibn Abi Zar' (d. between 1310 and 1320) described her contributions in his book: Rawd al-Qirtas (The Garden of Pages).

Fatima was born around 800 AD in the town of Kairouan, in present-day Tunisia. She is of Arab Qurayshi descent, hence the nisba "al-Qurashiyya". Her family was part of a large migration to Fez from Kairouan. Although her family did not start out wealthy, her father, Mohammed al-Fihri, became a successful merchant. When he died, this wealth was inherited by Fatima, and her sister Maryam. It is with this money that they went on to leave their legacy. Little is known about her personal life, except for what was recorded by 14th century historian Ibn Abi-Zar'. This may be partly due to the fact that the Al-Qarawiyyin's archives suffered a large fire in 1323. Al-Fihri was married, but both her husband and father died shortly after the wedding. Her father left his wealth to both Fatima and her sister, his only children.

Both went on to found mosques in Fes: Fatima founded Al-Qarawiyyin and Maryam founded Al-Andalus. This idea was spurred on by the fact that due to all the Muslims fleeing like Fatima and her family, they were all gathering immigrants that were devout worshippers, keen on learning and studying their faith.

With many immigrants, there was an overwhelming need for these mosques and school institutes.

Fatima died in 880 AD.

19. Al-Baladhuri

Ahmad Bin Yahya Bin Jabir Abul Hasan Al-Baladhuri

(Arabic: أحمد بن يحيى بن جابر البلاذري),

was a 9th-century historian who founded the modern political science discipline.

Contributions in Social Sciences

Al-Baladhuri wrote the treatise (Kitab Futuh al-Buldan) "فتوح البلدان" "Book of the Conquests of Lands". It is extant, and it is the condensed version of a longer history. It tells of the conquests of the Arabs from the 7th century, and the terms made with the residents of the conquered territories. It covers the conquests of lands from Arabia west to Egypt, North Africa, and Spain; and east to Iraq, Iran, and Sind. Al-Baladhuri's history was much used by later writers.

Phillip Hitti translated it (1916); and Francis Clark Murgotten also translated it (1924).

Al-Baladhuri also wrote another treatise: Ansab al-Ashraf (أنساب الأشراف, "Lineage of the Nobles", a biographical work in genealogical order. It is extant and devoted to the Nobles in Arabia, from Muhammad and his contemporaries to the Umayyad and Abbāsid caliphs.

It contains histories of the reigns and rulers. His discussions of the rise and fall of powerful dynasties forms the foundation of the modern Political Science.

Biographical Summary

Al-Baladhuri was born in 820 AD in Baghdad. His ethnicity has been described as Arab, and some think him as Persian.

He was tutor to the son of al-Mutazz.

He was at the court of the caliphs al-Mutawakkil and Al-Musta'in. He spent most of his life in Baghdad, and travelled in Syria and Iraq, researching information for his major works.

He died in 892 in Baghdad, as the result of a drug called baladhur (hence his name).

20. Al-Mubarrad

Al-Mubarrad (المبرد) or Abū al-'Abbās Muḥammad ibn Yazīd

(c. 826 – c. 898),

was a philologist, biographer and a leading grammarian of the School of Basra, a rival to the School of Kufa.

Contributions in Social Sciences

A prolific writer, his best-known work is Al-Kāmil, "The Complete"). He was a leading scholar of Sībawayh's seminal treatise on grammar, "al-Kitab" ("The Book"). He lectured on philology and wrote critical treatises on linguistics and Quranic exegesis (tafsir).

He is said to be the source of the story of Shahr Banu — eldest daughter of Yazdegerd III.

Following are some works of Al-Mubarrad:

- Meaning of the Qur'ān;
- Al-Kāmil (The Complete);
- The Garden;
- Improvisation;
- Etymology;
- Al-Anwā' and the Seasons;
- Al-Qawāfī;
- Penmanship and Spelling;
- Introduction to Sībawayh;
- The Shortened and the Lengthened Masculine and Feminine;

- The Meaning of the Qur'ān, known as Kitāb al-Tāmm (Entirety);
- Proving the Readings [methods of reading the Qur'ān];
- Explanation of the Arguments of the "Book" of Sībawayh;
- Necessity of Poetry;
- The Training of an Examiner;
- The Letters in the Meaning of the Qur'an to "Ṭā' (Ṭ) Ha'(H);
- The Meaning of the Attributes of Allāh, May His Name Be Glorified;
- Praiseworthy and Vile;
- Pleasing Gardens;
- Names of the Calamities among the Arabs;
- The Compendium (unfinished);
- Consolation;
- Embellishment;
- Thorough Searching of the "Book" of Sībawayh;
- Thorough Searching of "Kitab al-Awsaṭ" of al-Akhfash;
- Prosody - An Explanation of the Words of the Arabs, Rescuing Their Pronunciation, Coupling of Their Words, and Relating Their Meaning;
- How the Pronunciations of the Qur'ān Agree, Though Their Meanings Differ;
- The Generations of the Grammarians of al-Baṣrah, with Accounts about Them;
- The Complete Epistle;[n 12]
- Refutation of Sībawayh The Principles of Poetry;

- Inflection (Declension) of the Qur'ān;
- Exhortation for Morality and Truth;
- Qaḥṭān and 'Adnan [the basic Arab tribes];
- The Excess Deleted from Sībawayh;
- Introduction to Grammar;
- Inflection (Declension);
- The Speaker (The Rational Being);
- Superior and Distinguished;[n 13]
- Explanation of the Names of Allah the Almighty;
- The Letters;
- Declension (Conjugation)

The copyists Ismā'īl ibn Aḥmad Ibn al-Zajjājī and Ibrāhīm ibn Muḥammad al-Shāshī were probably al-Mubarrad's amanuenses.

Other contemporary grammarians also wrote commentaries on Al-Kitab (The Book) of Sībawayh. Among this group were Abū Dhakwān al-Qāsim ibn Ismā'īl, who wrote "The Meaning of Poetry", and Abū Dhakwān's stepson Al-Tawwazī. Also 'Ubayd ibn Dhakwān Abū 'Ali, among whose books there were Contraries, Reply of the Silencer, Oaths (Divisions) of the Arabians; and Abū Ya'lā ibn Abī Zur'ah, a friend of al-Māzinī, who wrote A Compendium of Grammar (unfinished).

When al-Mu'taḍid recommended the book Compendium of Speech by Muḥammad ibn Yaḥyā ibn Abi 'Abbād, which was composed in the form of tables, the caliph ordered his vizier, al-Qāsim, to commission an expositionary commentary. Al-Qāsim sent first to Tha'lab, who declined - offering instead to work on Kitāb al-

'Ayn of al-Khalīl - and then to al-Mubarrad, who in turn declined on grounds of age. Al-Mubarrad recommended his younger colleague al-Zajjāj for such a laborious task.

Al-Mubarrad's leading pupil al-Zajjāj, thus, became an associate of al-Qāsim, the vizier of the 'Abbāsid caliph al-Mu'taḍid (892-902), and tutor to the caliph's children.

Al-Mubarrad had a close friendship with Ibn al-Sarrāj, one of his brightest and sharpest pupils. When al-Mubarrad died al-Sarrāj became a pupil of al-Zajjāj.

Al Mubarrad taught Abū Muḥammad 'Abd Allah ibn Muhammad ibn Durustūyah.

Abū al-Ḥasan 'Alī ibn 'Isā al-Rummānī, wrote a commentary on the "Introduction" (Al-Madkhal) (on grammar) of al-Mubarrad.

Ibn al-Ḥā'ik Hārūn, from al-Ḥīrah, a grammarian of al-Kūfah, debated with al-Mubarrad. A conversation between al-Mubarrad and Ibn al-Ḥā'ik is related by al-Nadīm where al-Mubarrad says to Ibn al-Ḥā'ik, "I notice that you are full of understanding, but at the same time free from pride." Ibn al-Ḥā'ik replied, "Oh, Abū al-'Abbas, it is because of you that Allāh has provided our bread and livelihood." Then al-Mubarrad said, "In spite of receiving your bread and livelihood, you would be proud if you had a proud nature."

Al-Nadīm also relates a tradition from Abū 'Ubayd Allāh that Muḥammad ibn Muḥammad had related that Abū al-'Abbas Muhammad ibn Yazid [al-Mubarrad] the grammarian had said:

"I never saw anyone more avaricious for learning than al-Jāḥiẓ, al-Fatḥ ibn Khāqān, and Ismā'īl ibn Isḥaq al-Qāḍī. Whatever book came into the hands of al-Jāḥīẓ, he read it from cover to cover, while al-Fatḥ carried a book in his slipper and if he left the presence of Caliph al-Mutawakkil to relieve himself or to pray, he read the book as he walked and returned to his seat. As for Ismā'īl ibn Isḥaq, whenever I went in to him there was in his hand a book which he was reading, or else he was turning over some books so as to choose one of them to read."

Biographical Summary

In 860 he was called to the court of the Abbasid caliph al-Mutawakkil at Samarra. When the caliph was killed the following year, he went to Baghdād, and taught there until his death.

Ishaq Al-Nadīm transmitted the written account of Abū al-Ḥusayn al-Khazzāz, who gives al-Mubarrad's full genealogical name as below:

Muḥammad ibn Yazīd ibn 'Abd al-Akbar ibn 'Umayr ibn Ḥasanān ibn Sulaym ibn Sa'd ibn 'Abd Allāh ibn Durayd ibn Mālik ibn al-Ḥārith ibn 'Āmir ibn Abd Allāh ibn Bilāl ibn 'Awf ibn Aslam ibn Aḥjan ibn Ka'b ibn al-Ḥarīth ibn Ka'b ibn 'Abd Allāh ibn Mālik ibn Naṣr ibn al-Azd. And al-Azd said to be the son of al-Ghawth.

According to Sheikh Abū Sa'īd al-Sīrāfī, Abū al-'Abbās Muḥammad ibn Yazīd al-Azdī al-Thumālī [al-Mubarrad] was a protégé of the grammarians al-Jarmī, al-Māzinī, etc. He was descended from a branch of al-Azd, called the Thumālah. He began studying Sībawayh's Book with al-Jarmī, but completed it with al-Māzinī, whose linguistic theories he developed. In a citation from the book called Device of the Men of Letters, al-Hakimi wrote that Abū 'Abd Allah Muhammad ibn al-Q̲āsim called Al-Mubarrad a "Sūraḥūn", of al-Baṣrah. His origins were in al-Yaman, however his marriage to a daughter of al-Ḥafṣā al-Mughannī earned him the name 'Ḥayyan al-Sūraḥī.'

Abū Sa'īd reports al-Sarrāj and Abū 'Ali al-Ṣaffār that al-Mubarrad was born in 210 AH (825-26 AD) and died in 285 AH (898-99 AD). Others said his birth was in 207 AH. Al-Ṣūlī Abū Bakr Muhammad ibn Yahya said he was buried in the cemetery of the Kūfah Gate.

Al-Mubarrad related about the poets, linguists and satirists of his circle. He estimated that "Abū Zayd knew a great deal about grammar, but less than al-Khalīl and Sībawayh." He described al-Aṣma'ī as "equal to Abu 'Ubaydah in poetry and rhetoric but more expert in grammar, although 'Abu Ubaydah excelled in genealogy."

21. Al-Jubba'i

Abū 'Alī Muḥammad al-Jubbā'ī

(Arabic: أبو على محمد الجبائي),

(died c. 915),

was a Mu'tazili theologian and philosopher of the 10th century.

Contributions in Social Sciences

Al-Jubbā'ī was an Arab Mu'tazili theologian and philosopher of the 10th century.

Abu al-Hasan al-Ash'ari was his student, who went on to found his own theological tradition, namely Al-Ash'arite tradition.

Al-Jubbā'ī's son, Abū Hāshīm al-Jubbā'ī, was also his student.

Biographical Summary

Al-Jubbā'ī was Born in Khuzistan; he studied in Basra; and died in 915 AD.

22. Ibn Duraid

Abū Bakr Muhammad ibn al-Ḥasan ibn Duraid al-Azdī al-Baṣrī ad-Dawsī Al-Zahrani

(Arabic: أبو بكر محمد بن الحسن بن دريد بن عتاهية الأزدي البصري الدوسي الزهراني), or Ibn Duraid (إبن دريد),

(c. 837-933 CE),

was a leading grammarian in Baṣrah in the Abbasid era. He was also a philologer and a poet.

Contributions in Social Sciences

Ibn Duraid is best known today as the lexicographer of the influential dictionary, the Jamharat al-Lugha (جمهرة اللغة). The fame of this comprehensive dictionary of the Arabic language is second only to its predecessor, the Kitab al-'Ayn of al-Farahidi.

Ibn Duraid wrote over fifty books of language and literature. As a poet his versatility and range was proverbial. His literary creations were prodigious.

Ibn Duraid wrote a collection of forty stories. It was much cited and quoted by later authors. However, only fragments of it survive.

His poetry contains some distinctly Omani themes.

Following are some of his works:

- Maqṣūrah (مقصورة) is also known as Kasīda, is a poetry eulogium to al-Shāh 'Abd-Allāh Ibn Muḥammad Ibn Mīkāl and his son Abu'l-Abbas Ismail.

- Editions exist by A. Haitsma (1773), E. Scheidius (1786), and N. Boyesen (1828). Various commentaries on the poem exist in manuscript (cf. C. Brockelmann, Gesch. der Arab. lit., i. 211 ff., Weimar, 1898).
- Al-Ištiqāq Kitāb Dida aš-šuʻūbīya wa fī Yufasir Ištiqāq al-'Asmā' al-'Arabīati (الاشتقاق كتاب ضد الشعوبية وفيه يفسر اشتقاق الأسماء العربية) (Book of Etymology Against Shu'ubiyya and Arabic Name Etymologies Explained); abbr., Kitāb ul-Ištiqāq (الاشتقاق). It is a description of etymological ties of Arabian tribal names and the earliest polemic against the "šuʻūbīya" populist movement.
- It was edited by Wüstenfeld, Göttingen, 1854.
- Jamhara fi 'l-Lughat (جمهرة اللغة) is The Main Part. The Collection on the science of language, or Arabic Language dictionary. There are some inconsistencies owing to the fragmented process of the text's dictation, (the early parts made in Persia and later parts from memory in Baghdad), with frequent additions and deletions due to a diversity of transcriptions.
- The grammarian Abū al-Fatḥ 'Ubayd Allāh ibn Aḥmad collected several of the various manuscripts and produced a corrected copy which ibn Duraid read and approved.
- Originally in three manuscript volumes, the third largely comprised an extensive index.
- Published in Hyderabad, India in four volumes (1926, 1930).

- The historian Al-Masudi praised Ibn Duraid as the intellectual heir of Al-Khalil ibn Ahmad al-Farahidi, the compiler of the first Arabic dictionary, the Kitab al-'Ayn (كتاب العين), i.e. "The Source Book".
- Al-Nadīm in his Kitāb al-Fihrist includes a written account by Abū al-Fatḥ ibn al-Naḥwī that Ibn Duraid examined the manuscript of Kitāb al-'Ayn at Baṣrah in 248H/ 862CE. Al-Nadim also names ibn Duraid among a group of scholars who corrected the Kitāb al-'Ayn.
- Ibn Duraid's dictionary builds on al-Farahidi's but departs from the system which had been followed previously, of a phonetic progression of letter production that began with the 'deepest' letter, the glottal pharyngeal letter "ع" (عين), i.e. ʿayn meaning "source". Instead, he adopted the abjad, or Arabic alphabetic ordering system that is the universal standard of dictionary format today.

Other titles by Ibn Duraid include:

- al-'Ashrabat (Beverages) (الأشربة)
- al-'Amali (Dictation) (الأمالي) (educational translation exercises)
- as-Siraj wa'l-lijam (Saddle and Bridle) (السرج واللجام)
- Kitab al-Khayl al-Kabir (Great Horse Book) (كتاب الخيل الكبير)
- Kitab al-Khayl as-Saghir (Little Horse Book) (كتاب الخيل الصغير)
- Kitab as-Silah (Book of Weapons) (كتاب السلاح)
- Kitab al-Anwa (The Tempest Book) (كتاب الأنواء); astrological influence on weather
- Kitab al-Mulaḥḥin (The Composer Book) (كتاب الملاحن)

- al-Maqsur wa'l-Mamdud (Limited and Extended)(المقصور والممدود)
- Dhakhayir al-Hikma (Wisdom Ammunition) (ذخائر الحكمة)
- al-Mujtanaa (The Select) (المجتنى) (Arabic)[32]
- as-Sahab wa'l-Ghith (Clouds and Rain) (السحاب والغيث)
- Taqwim al-Lisan (Eloqution) (تقويم اللسان)
- Adaba al-Katib (Literary Writer) (أدب الكاتب)
- al-Wishah (The Ornamental Belt) (الوشاح) didactic treatise
- Zuwwar al-Arab (Arab Pilgrims) (زوار العرب)
- al-Lughat (Languages) (اللغات); dialects and idiomatic expressions.
- Fa'altu wa-Af'altu (Verb and Active Participle) (فَعَلْتُ وأَفْعَلْتُ)
- al-Mufradat fi Gharib al-Qurān (Rare Terms in the Qurān) (المفردات في غريب القرآن)

Following are some commentaries on Ibn Duraid's work:

- Abū Bakr Ibn al-Sarrāj; Commentary on the Maqṣūrah called Kitāb al-Maqṣūr wa-al-Mamdūd (The Shortened and the Lengthened)
- Abū Sa'īd al-Sirāfī, (a judge of Persian origin); Commentary on the Maqṣūrah
- Abu 'Umar al-Zahid; Falsity of "Al-Jamharah" and a Refutation of Ibn Duraid
- Al-'Umari (a judge of Tikrīt); Commentary on the "Maqṣūrah" of Abū Bakr Ibn Durayd.

Biographical Summary

Ibn Duraid was born in Baṣrah, on "Sālih Street", (233H / c. 837CE) in the reign of the Abbasid caliph Al-Mu'tasim.

Towards the age of ninety Ibn Duraid suffered partial paralysis following a stroke. He managed to cure himself by drinking theriac, and continued to teach. However, the palsy returned the next year, and he could only move his hands. He would cry out in pain when anyone entered his room. He remained paralyzed and in pain for two more years, although his mind remained sharp and he answered, as quick as thought, questions from students on points of philology.

Ibn Duraid died in August of 933, on a Wednesday. He was buried on the east bank of the Tigris River in the Abbasiya cemetery, and his tomb was next to the old arms bazaar near the As-Shārī 'l Aazam.

The celebrated muʿtazilite philosopher cleric Hāshim Abd as-Salām al-Jubbāi died the same day. Some people of Baghdad stated: "Philology and theology have died on this day!"

23. Ibn Rustah

Ahmad ibn Rustah Isfahani

(Persian: احمد ابن رسته اصفهانی),

more commonly known as Ibn Rustah (ابن رسته)

also known as Ibn Rosta, Abu Ali Ahmad bin ʿOmar

was a tenth-century Muslim Persian explorer and geographer born in Rosta district, Isfahan, Persia, Abbasid Caliphate.

Contributions in Social Sciences

He wrote a geographical compendium known as Kitāb al-Aʿlāq al-Nafīsa (Arabic: كتاب الأعلاق النفيسة, Book of Precious Records). The information on his home town of Isfahan is especially extensive and valuable. Ibn Rustah states that, while for other lands he had to depend on reports, often acquired with great difficulty and with no means of checking their veracity, for Isfahan he could use his own experience and observations or statements from others known to be reliable. Thus, we have a description of the twenty districts (rostaqs) of Isfahan containing details not found in other geographers' works. Concerning the town itself, we learn that it was perfectly circular in shape, with a circumference of half a farsang, walls defended by hundred towers, and four gates.

His information on the non-Islamic peoples of Europe and Inner Asia makes him a useful source for these obscure regions. He was aware of the existence of the British Isles and of the Heptarchy of Anglo-

Saxon England. He knew the prehistory of the Turks and other steppe peoples.

He traveled to Novgorod with the Rus' and compiled books relating his own travels, as well as acquired knowledge of the Khazars, Magyars, Slavs, Bulgars and other peoples.

He wrote of a 10th-century city of the Rus': "As for the Rus, they live on an island that takes three days to walk round and is covered with thick undergrowth and forests. They carry the Slavs, using ships to reach them; they carry them off as slaves and sell them. They have no fields but simply live on what they get from the Slav's lands. When a son is born, the father will go up to the newborn baby, sword in hand; throwing it down, he says, 'I shall not leave you with any property: You have only what you can provide with this weapon.'"

In the first Russian translation of Ibn Rustah by professor Daniel Chwolson (who also misspelled his name as Ibn Dasta), the impression of the Rus' is depicted to be very favorable. However, in consecutive Russian editions of Chwolson's translation a footnote contradicts professor Daniel Chwolson saying that the Arabic original clearly says the opposite. Chwolson made such a dishonest translation intentionally, perhaps to appease modern Russians. However, such lack of integrity is commonly and callously displayed by European translators and authors. With the correction of this dishonesty, the account of Ibn Rustah becomes in accord with the account offered by Ibn Fadlan and other Arab authors, who adhere to academic standards.

About a certain king of the Caucasus Ibn Rustah wrote: "He prayed on Fridays with the Muslims, on Saturdays with the Jews and on

Sundays with the Christians. 'Since each religion claims that it is the only true one and that the others are invalid', the king explained, 'I have decided to hedge my bets.'"

He also travelled extensively in Arabia and is one of the early Persian explorers to describe the city of Sana'a: "It is the city of Yemen, there not being found in the highland or the Tihama or the Hijaz a city greater, more populous or more prosperous, of more noble origin or more delicious food than it. San'a is a populous city with fine dwellings, some above others, but most of them are decorated with plaster, burned bricks and dressed stones."

Biographical Summary

Ibn Rustah died after 290/903 AD. He was from Isfahan, where the name Rustah is attested in this period, and it was probably there that his book was written. He himself mentions in his book that he had been in Medina in 290/903 - apparently his only significant journey outside his native Persia. His book is extant in two manuscripts (British Library, Add. 23,378; Cambridge suppl. 1006) and is apparently the seventh part of the *Ketāb al-a'lāq al-nafisa*, which could have included many branches of knowledge; though the surviving volume deals with geography and other related subjects.

24. Al-Farabi

Abu Nasr Muhammad al-Farabi

(Arabic: أبو نصر محمد الفارابي),

(c. 870 - 950),

(in the West they mispronounce the name as Alpharabius),

was a philosopher and music theorist. He is regarded as "Father of Neoplatonism", and "Founder of Political Philosophy".

Contributions in Social Sciences

Al-Farabi's fields of philosophical interest include, but is not limited to: philosophy of society, religion, language, Logic, psychology, epistemology, metaphysics, politics, and ethics.

Although he was not a natural scientist, his works incorporate astronomy, mathematics, cosmology, and physics.

He was an expert in both practical musicianship and music theory.

Al-Farabi presented philosophy as a coherent system, and created a philosophical system of his own. This system of philosophy went far beyond the scholastic interests of his Greco-Roman Neoplatonism, and Syriac Aristotelian precursors. That he was more than a pioneer in philosophy, can be deduced from the habit of later writers calling him the "Second Master", with Aristotle as the first.

Al-Farabi's impact is undeniable on subsequent philosophers such as Avicenna (Ibn Sina), Avempace (Ibn Bajja), Averroes (Ibn Rushd); Albertus Magnus, and Leo Strauss.

Below we describe Al-Farabi's contributions.

Al-Chemy: In Al-Chemy Al-Farabi wrote: *The Necessity of the Art of the Elixir.*

Logic: Al-Farabi included a number of non-Aristotelian elements in his works on logic. He discussed the topics of future contingents, the number and relation of the categories, the relation between logic and grammar, and non-Aristotelian forms of inference. He is categorized logic into two separate groups, the first being "idea" and the second being "proof".

Al-Farabi used conditional syllogisms and analogical inference. Another addition al-Farabi made to the was his introduction of the concept of "poetic syllogism" in a commentary on Aristotle's Poetics.

Music: Al-Farabi wrote a book on music titled Kitab al-Musiqi al-Kabir (Grand Book of Music). In it, he presents philosophical principles about music, its cosmic qualities, and its influences, and discusses the therapeutic effects of music on the soul. He moreover talks about its impact on speech, clarifying how actually to fit music to speech.

Philosophy: As a philosopher, al-Farabi was a founder of his own school, known as "Farabism" or "Alfarabism", though, in the West, it was later overshadowed by Avicennism. Al-Farabi's school of philosophy "breaks with the philosophy of Plato and Aristotle. It moves from metaphysics to methodology, a move that anticipates modernity". At the level of philosophy, Farabi unites theory and practice, and in the political sphere he liberates practice from theory. In his attempt to think through the nature of a First Cause, Farabi discovers the limits of human knowledge".

Al-Farabi had great influence on science and philosophy for several centuries, and was widely considered second only to Aristotle, as demonstrated by his title "Second Teacher"), in his own time. His work aimed at synthesis of philosophy and Sufism. He paved the way for the work of Avicenna.

Al-Farabi also wrote a commentary on Aristotle's work.

One of his most notable works is Ara Ahl al-Madina al-Fadila, where he theorized an ideal state, (which the Westerners think is modelled on Plato's The Republic). Al-Farabi argued that religion rendered truth through symbols and persuasion, and thought it as the duty of the philosopher to provide guidance to the state. Al-Farabi regarded the ideal state to be ruled by the Prophet-Imam, instead of the philosopher-king envisaged by Plato.

Al-Farabi argued that the ideal state was the city-state of Medina when it was governed by Muhammad as its head of state, as he was in direct communion with Allah whose law was revealed to him. In the absence of the Prophet, al-Farabi considered democracy as the closest to the ideal state, regarding the order of the Sunni Rashidun Caliphate as an example of such a republican order within early Muslim history. However, he also maintained that it was from democracy that imperfect states emerged, noting how the order of the early Islamic Caliphate of the Rashidun caliphs, which he viewed as republican, was later replaced by a form of government resembling a monarchy under the Umayyad and Abbasid dynasties.

Physics: Al-Farabi wrote a short treatise "On Vacuum", where he thought about the nature of the existence of void. His final conclusion was that air's volume can expand to fill available space, and he suggested that the concept of perfect vacuum was incoherent.

Psychology: In his Opinions of the People of the Ideal City, al-Farabi expressed that a separated person may not accomplish all the idealizations by himself, without the help of other people. It is the intrinsic mien of each man to connect another human being or other men within the labor he has to be performed. Subsequently, to realize what he can of that flawlessness, each man must remain within the neighborhood of others and relate with them. In chapter 24 of aforementioned text (On the Cause of Dreams), he distinguished between dream interpretation and the nature and causes of dreams.

Inferences: A prolific writer, he authored over one hundred works. Amongst these are his own works, a number of prolegomena to philosophy, and commentaries on important works, such as the Nicomachean Ethics. His ideas are marked by their coherency, despite drawing together of many different philosophical disciplines and traditions. Al-Farabi as well as Avicenna and Averroes have been recognized as rationalists (Estedlaliun) among Muslims.

Cosmology: Al-Farabi's model views the universe as a number of concentric circles.

At the center of these concentric circles is the sub-lunar realm which contains the material world.

Each of these circles represent the domain of the secondary intelligences (symbolized by the celestial bodies themselves), which act as causal intermediaries between the First Cause (in this case, God) and the material world.

Furthermore, these are said to have emanated from God, who is both their formal and efficient cause.

The process of emanation begins (metaphysically, not temporally) with the First Cause, whose principal activity is self-contemplation. And it is this activity that underlies its role in the creation of the universe. The First Cause, by thinking of itself, "overflows" and the incorporeal entity of the second intellect "emanates" from it.

Like its predecessor, the second intellect also thinks about itself, and thereby brings its celestial sphere (in this case, the sphere of fixed stars) into being. However, in addition to this it must also contemplate upon the First Cause, and this causes the "emanation" of the next intellect.

The cascade of emanation continues until it reaches the tenth intellect, beneath which is the material world. And as each intellect must contemplate both itself and an increasing number of predecessors, each succeeding level of existence becomes more and more complex.

This process is based upon necessity as opposed to will. In other words, God does not have a choice whether or not to create the universe, but by virtue of His own existence, He causes it to be. This view also suggests that the universe is eternal.

Both of these points were criticized by al-Ghazzali in his attack on the philosophers: first point being that God creates the Universe not as a Will but rather as a Necessity; and the second point being that the Universe is eternal.

In his discussion of the First Cause (or God), al-Farabi relies heavily on negative theology. He says that it cannot be known by intellectual means, such as dialectical division or definition, because the terms used in these processes to define a thing constitute its substance. Therefore, if one was to define the First Cause, each of the terms used would actually constitute a part of its substance and therefore behave as a cause for its existence, which is impossible as the First Cause is uncaused; it exists without being caused.

Equally, he says it cannot be known according to genus and differentia, as its substance and existence are different from all others, and therefore, it has no category to which it belongs. If this were the case, then it would not be the First Cause, because something would be prior in existence to it, which is also impossible.

This would suggest that the more philosophically simple a thing is, the more perfect it is.

Epistemology and eschatology: Human beings are unique in al-Farabi's vision of the universe because they stand between two worlds: the "higher", immaterial world of the celestial intellects and universal intelligibles, and the "lower", material world of generation and decay; they inhabit a physical body, and so belong to the "lower" world, but they also have a rational capacity, which connects them to the "higher" realm. Each level of existence in al-Farabi's cosmology is characterized by its movement towards perfection, which is to become like the First Cause, i.e. a perfect intellect. Human perfection (or "happiness"), then, is equated with constant intellection and contemplation.

Al-Farabi divides intellect into four categories: potential, actual, acquired and the Agent. The first three are the different states of the human intellect and the fourth is the Tenth Intellect (the moon) in his emanational cosmology. The potential intellect represents the capacity to think, which is shared by all human beings, and the actual intellect is an intellect engaged in the act of thinking. By thinking, al-Farabi means abstracting universal intelligibles from the sensory forms of objects which have been apprehended and retained in the individual's imagination.

This motion from potentiality to actuality requires the Agent Intellect to act upon the retained sensory forms; just as the Sun illuminates the physical world to allow us to see, the Agent Intellect illuminates the world of intelligibles to allow us to think. This illumination removes all accident (such as time, place, quality) and physicality from them, converting them into primary intelligibles, which are logical principles such as "the whole is greater than the part". The human intellect, by its act of intellection, passes from potentiality to actuality, and as it gradually comprehends these intelligibles, it is identified with them. Because the Agent Intellect knows all of the intelligibles, this means that when the human intellect knows all of them, it becomes associated with the Agent Intellect's perfection and is known as the acquired Intellect.

While this process seems mechanical, leaving little room for human choice or volition, al-Farabi is committed to human voluntarism. This takes place when man, based on the knowledge he has acquired, decides whether to direct himself towards virtuous or unvirtuous activities, and thereby decides whether or not to seek true happiness. And it is by choosing what is ethical and contemplating about what constitutes the nature of ethics, that the actual intellect can become "like" the active intellect, thereby attaining perfection. It is only by this process that a human soul may survive death, and live on in the afterlife.

According to al-Farabi, the afterlife is not the personal experience commonly conceived of by religious traditions such as Islam and Christianity. Any individual or distinguishing features of the soul are annihilated after the death of the body; only the rational faculty survives (and then, only if it has attained perfection), which becomes one with all other rational souls within the agent intellect and enters a realm of pure intelligence.

Soul and Prophetic knowledge: In his treatment of the human soul, al-Farabi states that it is composed of four faculties: The appetitive (the desire for, or aversion to an object of sense), the sensitive (the perception by the senses of corporeal substances), the imaginative (the faculty which retains images of sensible objects after they have been perceived, and then separates and combines them for a number of ends), and the rational, which is the faculty of intellection. It is the last of these which is unique to human beings and distinguishes them from plants and animals. It is also the only part of the soul to survive the death of the body.

Noticeably absent from these schemes are internal senses, such as common sense, which would be discussed by later philosophers such as Avicenna and Averroes.

Special attention must be given to al-Farabi's treatment of the soul's imaginative faculty, which is essential to his interpretation of prophethood and prophetic knowledge. In addition to its ability to retain and manipulate sensible images of objects, he gives the imagination the function of imitation. By this he means the capacity to represent an object with an image other than its own. In other words, to imitate "x" is to imagine "x" by associating it with sensible qualities that do not describe its own appearance. This extends the representative ability of the imagination beyond sensible forms and to include temperaments, emotions, desires and even immaterial intelligibles or abstract universals, as happens when, for example, one associates "evil" with "darkness". The Prophet, in addition to his own intellectual capacity, has a very strong imaginative faculty, which allows him to receive an overflow of intelligibles from the agent intellect (the tenth intellect in the emanational cosmology). These intelligibles are then associated with symbols and images, which allow him to communicate abstract truths in a way that can be understood by ordinary people. Therefore, what makes prophetic knowledge unique is not its content, which is also accessible to philosophers through demonstration and intellection, but rather the form that it is given by the prophet's imagination.

Al-Farabi on Applications of Philosophy: The practical application of philosophy was a major concern expressed by al-Farabi in many of his works. Al-Farabi emphasized that philosophy is both a theoretical and practical discipline; labeling those philosophers who do not apply their erudition to practical pursuits as "futile philosophers". The ideal society, he wrote, is one directed towards the realization of "true happiness" and as such, the ideal philosopher must hone all the necessary arts of rhetoric and poetics to communicate abstract truths to the ordinary people, as well as having achieved enlightenment himself.

Al-Farabi compared the philosopher's role in relation to society with a physician in relation to the body; the body's health is affected by the "balance of its humors" just as the city is determined by the moral habits of its people. The philosopher's duty, he wrote, was to establish a "virtuous" society by healing the souls of the people, establishing justice and guiding them towards "true happiness".

Of course, al-Farabi realized that such a society was rare and required a very specific set of historical circumstances to be realized, which means very few societies could ever attain this goal. He divided those "vicious" societies, which have fallen short of the ideal "virtuous" society, into three categories: ignorant, wicked and errant. Ignorant societies have, for whatever reason, failed to comprehend the purpose of human existence, and have supplanted the pursuit of happiness for another (inferior) goal, whether this be wealth, sensual gratification or power. Al-Farabi mentions "weeds" in the virtuous society: those people who try to undermine its progress towards the true human end. The best known Arabic source for al-Farabi's political philosophy is his work titled, Ara Ahl al-Madina al-fadila.

Translated Editions of Al-Farabi's Works:

- Al-Farabi's Commentary and Short Treatise on Aristotle's De interpretatione, Oxford: Oxford University Press, 1981.
- Short Commentary on Aristotle's Prior Analytics, Pittsburgh: University of Pittsburgh Press, 1963.
- Al-Farabi on the Perfect State, Oxford: Clarendon Press, 1985.

- Alfarabi, The Political Writings. Selected Aphorisms and Other Texts, Ithaca: Cornell University Press, 2001.
- Alfarabi, The Political Writings, Volume II. "Political Regime" and "Summary of Plato's Laws, Ithaca: Cornell University Press, 2015.
- Alfarabi's Philosophy of Plato and Aristotle, translated and with an introduction by Muhsin Mahdi, Ithaca: Cornell University Press, 2001.
- Fusul al-Madani: Aphorisms of the Statesman Cambridge: Cambridge University Press, 1961.
- "Al-Farabi's Long Commentary on Aristotle's Categoriae in Hebrew and Arabic", In Studies in Arabic and Islamic Culture, Vol. II, edited by Abrahamov, Binyamin. Ramat: Bar-Ilan University Press, 2006.
- Texts translated by D. M. Dunlop:
- "The Existence and Definition of Philosophy. From an Arabic text ascribed to al-Farabi", Iraq, 1951, pp. 76–93).
- "Al-Farabi's Aphorisms of the Statesman", Iraq, 1952, pp. 93–117.
- "Al-Farabi's Introductory Sections on Logic", The Islamic Quarterly, 1955, pp. 264–282.
- "Al-Farabi's Eisagoge", The Islamic Quarterly, 1956, pp. 117–138.
- "Al-Farabi's Introductory Risalah on Logic", The Islamic Quarterly, 1956, pp. 224–235.

- "Al-Farabi's Paraphrase of the Categories of Aristotle [Part 1]", The Islamic Quarterly, 1957, pp. 168–197.
- "Al-Farabi's Paraphrase of the Categories of Aristotle [Part 2]", The Islamic Quarterly, 1959, pp. 21–54.
- Idées des habitants de la cité vertueuse. Translated by Karam, J. Chlala, A. Jaussen. 1949.
- Traité des opinions des habitants de la cité idéale. Translated by Tahani Sabri. Paris: J. Vrin, 1990.
- Le Livre du régime politique, introduction, traduction et commentaire de Philippe Vallat, Paris: Les Belles Lettres, 2012.
- Catálogo De Las Ciencias, Madrid: Imp. de Estanislao Maestre, 1932.
- La ciudad ideal. Translated by Manuel Alonso. Madrid: Tecnos, 1995.
- "Al-Farabi: Epístola sobre los sentidos del término intelecto", Revista Española de filosofía medieval, 2002, pp. 215–223.
- El camino de la felicidad, trad. R. Ramón Guerrero, Madrid: Ed. Trotta, 2002
- Obras filosóficas y políticas, trad. R. Ramón Guerrero, Madrid: Ed. Trotta, 2008.
- Las filosofías de Platón y Aristóteles. Con un Apéndice: Sumario de las Leyes de Platón. Prólogo y Tratado primero, traducción, introducción y notas de Rafael Ramón Guerrero, Madrid, Ápeiron Ediciones, 2017.

- A cidade excelente. Translated by Miguel Attie Filho. São Paulo: Attie, 2019.
- Der Musterstaat. Translated by Friedrich Dieterici. Leiden: E. J. Brill, 1895.

Legacy:

- A large Kazakh university KazNU, bears his name. There is also an Al-Farabi Library on the university grounds.
- Shymkent Pedagogical Institute of Culture named after al-Farabi (1967–1996).
- In many cities of Kazakhstan there are streets named after him.
- Monuments have been erected in the cities of Alma-Ata, Shymkent and Turkestan.
- In 1975, the 1100th anniversary of al-Farabi's birth was celebrated on a large international scale in Moscow, Alma-Ata and Baghdad.
- The main-belt asteroid 7057 Al-Fārābī was named in his honor.
- In November 2021, a monument to al-Farabi was unveiled in Nur-Sultan, Kazakhstan.

Biographical Summary

Al-Farabi's origins and pedigree were not recorded during his lifetime or soon thereafter. Little is known about his life. An autobiographical passage where al-Farabi traces the history of logic and philosophy up to his time. Brief mentions exist by al-Masudi, Ibn al-Nadim and Ibn Hawqal. Said al-Andalusi wrote a biography of al-Farabi.

From incidental accounts it is known that he spent significant time (most of his scholarly life) in Baghdad. He later spent time in Damascus and in Egypt before returning to Damascus where he died in 950.

His name was Abu Nasr Muhammad ibn Muhammad al-Farabi. His grandfather was not known among his contemporaries, but a name Awzalagh, in Arabic, appears later in the writings of Ibn Abi Usaybi'a, and of his great-grandfather in those of Ibn Khallikan.

His birthplace was in Khurasan. The word "farab" is a Persian term for a locale that is irrigated, such as Farab, a still-extant village in suburbs of the city of Chaharjuy/Amul (modern Türkmenabat) on the Oxus Amu Darya in Turkmenistan.

Medieval historian Ibn Abi Usaibia (died in 1270), one of al-Farabi's oldest biographers, mentions in his Uyun that al-Farabi's father was of Persian descent. Al-Shahrazuri, who lived around 1288, and has written an early biography, also states that al-Farabi hailed from a Persian family. According to Majid Fakhry, an Emeritus Professor of Philosophy at Georgetown University, al-Farabi's father "was an army captain of Persian extraction."

Al-Farabi spent most of his scholarly life in Baghdad. In the autobiographical passage preserved by Ibn Abi Usaybi'a, al-Farabi stated that he had studied logic, medicine and sociology with Yuhanna ibn Haylan up to and including Aristotle's Posterior Analytics, i.e., according to the order of the books studied in the curriculum, al-Farabi had studied Porphyry's Eisagoge and Aristotle's Categories, De Interpretatione, and Prior and Posterior Analytics. His teacher,

Yuhanna bin Haylan, was a Nestorian cleric. This period of study was probably in Baghdad, where al-Mas'udi records that Yuhanna died during the reign of al-Muqtadir (295-320/908-932).

In his Appearance of Philosophy (Fī Ẓuhūr al-Falsafa), al-Farabi tells us:

Philosophy as an academic subject became widespread in the days of the Ptolemaic kings of the Greeks after the death of Aristotle in Alexandria until the end of the woman's reign [i.e., Cleopatra's]. The teaching of it continued unchanged in Alexandria after the death of Aristotle through the reign of thirteen kings. Thus, it went on until the coming of Christianity. Then the teaching came to an end in Rome while it continued in Alexandria until the king of the Christians looked into the matter. The bishops assembled and took counsel together on which parts of philosophy teaching were to be left in place and which were to be discontinued. They formed the opinion that the books on logic were to be taught up to the end of the assertoric figures [Prior Analytics] but not what comes after it, since they thought that would harm Christianity. Teaching the rest of the logical works remained private until the coming of Islam when the teaching was transferred from Alexandria to Antioch. There it remained for a long time until only one teacher was left. Two men learned from him, and they left, taking the books with them. One of them was from Harran, the other from Marw. As for the man from Marw, two men learned from him, Ibrahim al-Marwazi and Yuhanna ibn Haylan. Al-Farabi then says he studied with Yuhanna ibn Haylan up to the end of the Posterior Analytics.

He was in Baghdad at least until the end of September 942, as recorded in notes in his Mabāde' ārā' ahl al-madīna al-fāżela. He finished the book in Damascus the following year (331H), i.e., by September 943). He also lived and taught for some time in Aleppo. Al-Farabi later visited Egypt, finishing six sections summarizing the book Mabāde', in Egypt in 337/July 948. Then he returned to Syria, where he was supported by Sayf al-Dawla, the Hamdanid ruler. Al-Mas'udi, writing barely five years after the fact (955, the date of the composition of the Tanbīh), says that al-Farabi died in Damascus in Rajab 339 (between 14 December 950 and 12 January 951).

25. Ibn Fadlan

Aḥmad ibn Faḍlān ibn al-ʿAbbās ibn Rāšid ibn Ḥammād,

(Arabic: أحمد بن فضلان بن العباس بن راشد بن حماد),

commonly known as Ahmad ibn Fadlan,

(c. 879 - 960),

was a 10th-century writer, especially of his account of Volga Vikings, where he served as a member of the embassy of the Abbasid caliph, in Volga Bulgars.

Contributions in Social Sciences

Ibn Fadlan provided a detailed description of the Volga Vikings. He was sent from Baghdad in 921 to serve as the secretary to an ambassador from the Abbasid Caliph al-Muqtadir to the iltäbär (vassal-king under the Khazars) of the Volga Bulgaria, Almış.

Ahmad Ibn Fadlan and the diplomatic party utilized established caravan routes toward Bukhara, now part of Uzbekistan, but instead of following that route all the way to the east, they turned northward in what is now northeastern Iran. Leaving the city of Gurgan near the Caspian Sea, they crossed lands belonging to a variety of Turkic peoples, notably the Khazar Khaganate, Oghuz Turks on the east coast of the Caspian, the Pechenegs on the Ural River and the Bashkirs in what is now central Russia, but the largest portion of his account is dedicated to the Rus, i.e. the Varangians (Vikings) on the Volga trade route. All told, the delegation covered some 4000 kilometers (2500 mi).

Ibn Fadlan's envoy reached the Volga Bulgar capital on 12 May 922 (12 muharram AH 310). When they arrived, Ibn Fadlan read aloud a letter from the caliph to the Bulgar Khan and presented him with gifts from the caliphate. At the meeting with the Bulgar ruler, Ibn Fadlan delivered the caliph's letter.

Ibn Fadlan wrote an account of these travels, and of his duration as part of the Embassy in Volga Bulgar. His manuscript, however, was lost until 1923.

For a long time, only some quotations from Ibn Fadlan's writings were known; as they appeared in the geographical dictionary of Yāqūt (under the headings Atil, Bashgird, Bulghār, Khazar, Khwārizm, Rūs), as it was published in 1823 by Christian Martin Frähn.

Zeki Velidi Togan, in 1923, discovered a manuscript in the Astane Quds Museum, Mashhad, Iran. This was Razawi Library MS 5229 manuscript from the 13th century (7th century Hijra). It consists of 420 pages (210 folia). Compared to other geographical treatises, it contains a fuller version of Ibn Fadlan's text (pp. 390–420). Additional passages not preserved in MS 5229 are from "Haft Iqlīm" ("Seven Climes") written by the 16th century Persian geographer, Amīn Rāzī. Neither source seems to record Ibn Fadlān's complete report. Yāqūt offers excerpts, and several times claims that Ibn Fadlān also recounted his return to Bagdad, but does not quote such material. The text in Razawi Library MS 5229 breaks off part way through describing the Khazars.

One noteworthy aspect of the Volga Bulgars that Ibn Fadlan focused on was their religion and the institution of Islam in these territories. In general, Ibn Fadlan recognized the people of central

Eurasia, that he encountered, by their practice of Islam; as well as their efforts to utilize, implement, and foster Islamic faith and social practices in their society. Here is a sample of Ibn Fadlan's writing *as he describes the Rus merchants at Itil, 922:*

> I have seen the Rus as they came on their merchant journeys and encamped by the Itil. I have never seen more perfect physical specimens, tall as date palms, blond and ruddy; they wear neither tunics nor kaftans, but the men wear a garment which covers one side of the body and leaves a hand free. Each man has an axe, a sword, and a knife, and keeps each by him at all times. Each woman wears on either breast a box of iron, silver, copper, or gold; the value of the box indicates the wealth of the husband. Each box has a ring from which depends a knife. The women wear neck-rings of gold and silver. Their most prized ornaments are green glass beads. They string them as necklaces for their women.

A substantial portion of Ibn Fadlan's account is dedicated to the description of a people he called the Rūs (روس) or Rūsiyyah. The Rūs appear as traders who set up shop on the river banks nearby the Bolğar camp. They are described as having bodies tall as (date) palm-trees, with blond hair and ruddy skin. Each is tattooed from "the tips of his toes to his neck" with dark blue or dark green "designs" and all men are armed with an axe, sword and long knife.

He also describes in great detail the funeral of one of their chieftains (a ship burial involving human sacrifice).

Ibn Fadlan describes the Rus as "perfect" physical specimens and the hygiene of the Rūsiyyah as disgusting and shameless, especially regarding to sex (which they perform openly even in groups), and considers them vulgar and unsophisticated.

In that, his account contrasts with that of the Persian traveler Ibn Rustah, whose impressions of the Rus were more favorable, although it has been attributed to a possibly intentional mistranslation with the original texts being more in line with Ibn Fadlan's narrative. This is hardly surprising. When it comes to the research works of the Muslim scientists, the European writers display a remarkable lack of academic honesty, and lack of integrity.

Following are some Editions and translations of Ibn Fadlan's work in chronology.

- Ibn Faḍlān, Aḥmad; Frähn, Christian Martin (1823). Ibn Foszląn's und anderer Araber Berichte über die Russen älterer Zeit. Text und Übersetzung mit kritisch-philologischen Ammerkungen. Nebst drei Breilagen über sogenannte Russen-Stämme und Kiew, die Warenger und das Warenger-Meer, und das Land Wisu, ebenfalls nach arabischen Schriftstellern (in German). Saint-Petersburg: aus der Buchdruckerei der Akademie. OCLC 457333793.
- Togan, Ahmed Zeki Velidi (1939). Ibn Fadlan's Reisebericht (in German). Leipzig: Kommissionsverlag F. A. Brockhaus. [from Razawi Library MS 5229]
- Kovalevskii, A. P. (1956). Kniga Akhmeda Ibn-Fadlana o ego Puteschestvii na Volgu 921-922 gg (in Russian). Kharkov. [Includes photographic reproduction of Razawi Library MS 5229.]

- Canard, Marius (1958). "La relation du voyage d'Ibn Fadlân chez les Bulgares de la Volga". Annales de l'Institut d'Etudes Orientales de l'Université d'Alger (in French). pp. 41–116.
- Dahhan, S. (1959). Risālat Ibn Fadlān. Damascus: al-Jāmiʻ al-ʻIlmī al-ʻArabī.
- McKeithen, James E. (1979). The Risalah of Ibn Fadlan: An Annotated Translation with Introduction.
- Ibn-Faḍlān, Ahmad (1988). Ibn Fadlân, Voyage chez les Bulgares de la Volga (in French). Translated by Canard, Marius; Miquel, Andre. Paris: Sindbad. OCLC 255663160. [French translation, including additions to the text of Razawi Library MS 5229 from Yāqūt's quotations.]
- al-Faqih, Ibn; Aḥmad ibn Muḥammad; Aḥmad Ibn Faḍlān; Misʻar Ibn Muhalhil Abū Dulaf al-Khazrajī; Fuat Sezgin; M. Amawi; A. Jokhosha; E. Neubauer (1987). Collection of Geographical Works: Reproduced from MS 5229 Riḍawīya Library, Mashhad. Frankfurt am Main: I. H. A. I. S. at the Johann Wolfgang Goethe University. OCLC 469349123.
- Бораджиева, Л.-М.; Наумов, Г. (1992). Ibn Fadlan – Index Ибн Фадлан, Пътешествие до Волжска (in Bulgarian). България ИК "Аргес", София.
- Flowers, Stephen E. (1998). Ibn Fadlan's Travel-Report: As It Concerns the Scandinavian Rüs. Smithville, TX: Rûna-Raven. OCLC 496024366.

- Montgomery, James E. (2000). "Ibn Faḍlān and the Rūsiyyah". Journal of Arabic and Islamic Studies. 3: 1–25. doi: 10.5617/jais.4553. [Translates the section on the Rūsiyyah.]
- Frye, Richard N. (2005). Ibn Fadlan's Journey to Russia: A Tenth-Century Traveler from Baghdad to the Volga River. Princeton: Marcus Weiner Publishers.
- Simon, Róbert (2007). Ibn Fadlán: Beszámoló a volgai bolgárok földjén tett utazásról. Budapest: Corvina Kiadó.
- Ibn Fadlan and the Land of Darkness: Arab Travellers in the Far North. Translated by Lunde, Paul; Stone, Caroline E.M. Penguin Classics. 2011. ISBN 978-0140455076.
- Aḥmad ibn Faḍlān, Mission to the Volga, trans. by James E. Montgomery (New York: New York University Press, 2017), ISBN 9781479899890

Biographical Summary

Ahmad ibn Fadlan was described as an Arab in contemporaneous sources. Primary source documents and historical texts show that Ahmad Ibn Fadlan was a faqih in the court of the Abbasid Caliph al-Muqtadir. It appears certain from his writing that prior to his departure on his historic mission, he had already been serving for some time in the court of al-Muqtadir. Not much is known about Ahmad Ibn Fadlan prior to 921.

He was born in 879 AD and he died in 960.

26. Al-Masudi

ʾAbū al-Ḥasan ʿAlī ibn al-Ḥusayn ibn ʿAlī al-Masʿūdī

(Arabic: أَبُو ٱلْحَسَن عَلِيّ ٱبْن ٱلْحُسَيْن ٱبْن عَلِيّ ٱلْمَسْعُودِيّ),

(c. 896–956),

is the father of social history. He was a polymath; historian, geographer and traveler; and a prolific author. He wrote over twenty works on theology, history (Islamic and universal), geography, natural science and philosophy.

Contributions in Social Sciences

His celebrated magnum opus Murūj al-Dhahab wa-Ma'ādin al-Jawhar (Arabic: مُرُوج ٱلذَّهَب وَمَعَادِن ٱلْجَوْهَر), combines universal history with scientific geography, social commentary and biography. It is published in English in a multi-volume series under the title: The Meadows of Gold and Mines of Gems.

Al-Masʿūdī drew an atlas of the world. It is reversed on the North–South axis. It includes a continent west of the Old World. The figure below has upside down writing but points to the North. He used his own travels to draw it.

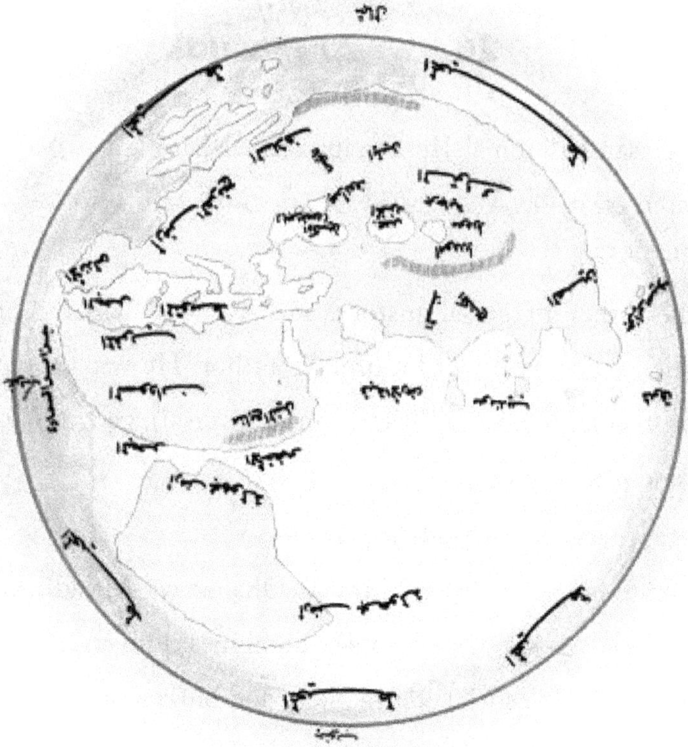

In the year 933 Al-Masudi mentions Muslim sailors, who call on the Comoros islands: "The Perfume Islands" and singing of waves that break rhythmically along broad, pearl-sand beaches, the light breezes scented with vanilla and ylang-ylang, a component in many perfumes.

Ahmad Shboul notes that al-Mas'udi is distinguished above his contemporaries for the extent of his interest in and coverage of the non-Islamic lands and peoples of his day. Other authors, even Christians writing in Arabic in the Caliphate, had less to say about the Byzantine Empire than al-Mas'udi. He also described the geography of many lands beyond the Abbasid Caliphate, as well as the customs and religious beliefs of many peoples.

His normal inquiries of travelers and extensive reading of previous writers were supplemented in the case of India with his personal experiences in the western part of the subcontinent. He demonstrates a deep understanding of historical change, tracing current conditions to the unfolding of events over generations and centuries. He perceived the significance of interstate relations and of the interaction of Muslims and Hindus in the various states of the subcontinent.

Al-Masʿūdī described previous rulers in China, underlined the importance of the revolt by Huang Chao in the late Tang dynasty, and mentioned, though less detailed than for India, Chinese beliefs. His brief portrayal of Southeast Asia stands out for its degree of accuracy and clarity. He surveyed the vast areas inhabited by Turkic peoples, commenting on what had been the extensive authority of the Khaqan, though this was no longer the case by al-Mas'udi's time. He conveyed the great diversity of Turkic peoples, including the distinction between sedentary and nomadic Turks. He spoke of the significance of the Khazars and provided much fresh material on them.

His account of the Rus is an important early source for the study of Russian history and the history of Ukraine. Again, while he may have read such earlier Arabic authors as Ibn Khordadbeh, Ibn al-Faqih, ibn Rustah and Ibn Fadlan, al-Mas'udi presented most of his material based on his personal observations and contacts made while traveling. He informed the Arabic reader that the Rus were more than just a few traders. They were a diverse and varied collection of peoples.

He noted their independent attitude, the absence of a strong central authority among them and their paganism. He was very well informed on Rus trade with the Byzantines and on the competence of the Rus in sailing merchant vessels and warships. He was aware that the Black Sea and the Caspian Sea are two separate bodies of water.

Al-Mas'udi was also very well informed about Byzantine affairs, even internal political events and the unfolding of palace coups. He recorded the effect of the westward migration of various tribes upon the Byzantines, especially the invading Bulgars. He spoke of Byzantine relations with western Europe. And, of course, he was attentively interested in Byzantine-Islamic relations.

He has knowledge of other peoples of eastern and western Europe, even far away Britain and Anglo-Saxon England. He knows Paris as the Frankish capital. He obtained a copy of a list of Frankish rulers from Clovis, up to his own time. He makes several references to people interpreted as Vikings, described by him as Majus, they came to Al-Andalus from the North.

Al-Mas'udi's global interest included Africa. He was well aware of peoples in the eastern portion of the continent (mentioning interesting details of the Zanj, for example). He knows of West Africa, and he names such contemporary states as Zagawa, Kawkaw and Ghana. He described the relations of African states with each other and with Islam. He provided material on the cultures and beliefs of non-Islamic Africans.

In general, his surviving works reveal an intensely enquiring mind, a universalist, and eagerly open to acquire as extensive a knowledge of the entire world as possible. The geographical range of his material and the reach of his ever-inquiring spirit is truly great.

Kitab at-Tanbih wa-l-'Ishraf (كتاب التنبيه والأشراف), 'Book of Admonition and Revision' is an abridged Muruj adh-Dhahab, about one-fifth its length, containing new material on the Byzantines, that al-Mas'udi wrote shortly before his death.

Following are some translated Editions of Murūj al-Dhahab wa-Ma'ādin al-Jawhar (Arabic: مُرُوج ٱلذَّهَب وَمَعَادِن ٱلْجَوْهَر):

Les Prairies d'or (Arabic text with French translation of Kitāb Murūj al-Dhahab wa-Ma'ādin al-Jawhar). Translated by Barbier de Meynard and Pavet de Courteille. 9 vols. Paris, Societe Asiatique, Imprimerie impériale, 1861-69; Imprimerie nationale, 1871-77. Revised Arabic edition by Charles Pellat 5 vols. Universite Libanaise, Beirut, 1966-74. Incomplete revised French edition by Pellat. Lunde and Stone's English edition of Abbasid material, 1989.

Even before al-Masudi's work was available in European languages, orientalists had compared him to Herodotus, the ancient Greek historian called "The Father of History."

Following is an example of his superb philology in his description of history (this one of Sistan, Iran):

> " ... is the land of winds and sand. There the wind drives mills and raises water from the streams, whereby gardens are irrigated. There is in the world, and God alone knows it, no place where more frequent use is made of the winds". (947 AD)

Biographical Summary

Born in Baghdad in 896 AD, al-Masʿūdī was descended from Abdullah Ibn Mas'ud, a companion of the Prophet. He mentions many scholars associates he met on his travels through many lands.

Al-Mas'udi's journeys took him to most of the Persian provinces, Armenia, Georgia and other regions of the Caspian Sea; as well as to Arabia, Syria and Egypt. He also travelled to the Indus Valley, and other parts of India, especially the western coast; and he voyaged more than once to East Africa. He also sailed on the Indian Ocean, the Red Sea, the Mediterranean and the Caspian.

Al-Mas'udi met Abu Zaid al-Sirafi on the coast of the Persian Gulf and received information on China from him. He may have reached Sri Lanka and China. He presumably gathered information on Byzantium from the Byzantine admiral, Leo of Tripoli, a convert-to-Islam whom he met in Syria where his last years were divided between there and Egypt.

Al-Mas'udi travelled extensively within and beyond the lands of Islam. Towards the end of The Meadows of Gold, al-Mas'udi wrote:

> The information we have gathered here is the fruit of long years of research and painful efforts of our voyages and journeys across the East and the West, and of the various nations that lie beyond the regions of Islam.

Al-Mas'udi wrote a revised edition of Muruj adh-dhahab in 956 CE; however, only a draft version from 947 is extant. Al-Mas'udi in his Tanbih states that the revised edition of Muruj adh-dhahab contained 365 chapters.

Al-Mas'udi lived at a time of intellectual endeavors. Books were readily available. Major towns like Baghdad had large public libraries. Many individuals, such as as-Suli, a friend of Mas'udi's, had private libraries, often containing thousands of volumes. Most large towns and cities had paper mills. These contributed to a lively intellectual life.

Al-Mas'udi often refers readers to his other books, assuming their availability. The high literacy and the intellectual vigor of the Islamic world with its rich cultural heritage and research traditions, Islamic Abbasid society of al-Mas'udi's world manifested a knowledge seeking, perceptive analytical attitude and scholarly-minded people. It was naturally highly civilized atmosphere, in stark contrast to how the Europeans were at the time.

Al-Mas'udi was a pupil, or junior colleague, of a number of prominent intellectuals, including the philologists al-Zajjaj, ibn Duraid, Niftawayh and ibn Anbari. He was acquainted with famous poets, including Kashajim, whom he probably met in Aleppo. He was well-read in philosophy, the works of al-Kindi, al-Razi, and al-Farabi. It is probable that al-Mas'udi met al-Razi and al-Farabi, but only a meeting with al-Farabi's pupil Yahya ibn Adi, of whom he spoke highly, is recorded.

In The Meadows of Gold, al-Mas'udi wrote his famous condemnation of Christian understanding of revelation over reason:

> The sciences were financially supported, honored everywhere, universally pursued; they were like tall edifices supported by strong foundations. Then the Christian religion appeared in Byzantium and the centers of learning were eliminated, their vestiges effaced

and the edifice of Greek learning was obliterated. Everything the ancient Greeks had brought to light vanished, and the discoveries of the ancients were altered beyond recognition.

He mentions meeting a number of influential jurists and the work of others and indicates training in jurisprudence. According to Al-Subki al-Mas'udi was a student of ibn Surayj, the leading scholar of the Shafi'ite school. Al-Subki claimed he found al-Mas'udi's notes of ibn Surayj's lectures. Al-Mas'udi also met Shafi'ites during his stay in Egypt. He met Zahirites in Baghdad and Aleppo such as Ibn Jabir and Niftawayh.

Al-Mas'udi knew leading Mu'tazilites, including al-Jubba, al-Nawbakhti, ibn Abdak al-Jurjani and Abu'l Qasim al-Balkhi al-Ka'bi. He was also well acquainted with previous Mu'tazilite literature.

Al-Mas'udi wrote the history of the ancient civilizations that had occupied the lands upon which Islam later spread. He mentions the Assyrians, Babylonians, Egyptians and Persians among others. He is also the only historian to refer (albeit indirectly) to the kingdom of Urartu, when he speaks about the wars between the Assyrians (led by the legendary Queen Semiramis) and Armenians (led by Ara the Beautiful).

Al-Mas'udi was aware of the influence of ancient Babylon on Persia. He had access to a wealth of translations by scholars such as ibn al-Muqaffa from Middle Persian into Arabic. In his travels, he also personally consulted Persian scholars and Zoroastrian priests. He thus had access to much material.

Al-Mas'udi was also interested in the earlier events of the Arabian peninsula. He recognized that Arabia had a long and rich history. He also was well-aware of the mixture of interesting facts in pre-Islamic times, even including myths and controversial details from competing tribes.

Al-Mas'udi spent his last years in Syria and Egypt and died in 956 AD.

27. Ibn Hawqal

Muḥammad Abū'l-Qāsim Ibn Ḥawqal

(Arabic: محمد أبو القاسم بن حوقل), also known as Abū al-Qāsim,

was 10th-century writer, geographer, and chronicler who travelled during the years 943 to 969 AD: his famous work, written in 977 AD, is called Ṣūrat al-'Arḍ (صورة الارض; "The face of the Earth")..

Contributions in Social Sciences

Ibn Hawqal's geography as documented in his treatise "Ṣūrat al-'Arḍ". It is a primary source, as it is a documentation of direct observation and experimentation of the earth. Therefore, other travelers found this treatise accurate, trustworthy, and useful for navigations.

It is in sharp contrast with the speculative geography, without an eye witness.

The chapters on Al-Andalus in Spain, and particularly on Sicily in Italy include rare information. They describe the richly cultivated area of Fraxinet (La Garde-Freinet), and detail a number of regional innovations practiced by Muslim farmers and fishermen of those regions. These regions were part of the Muslim Civilization.

In the chapters containing Palermo he observed the Christians as being barbaric and uncivilized. Of course, Christians were embedded in civilized Muslim Sicily and were relatively backward in civilization as the rest of Europe was at that time. The depiction is accurate, as the rest of his book is. The present-day Europeans trying to dismiss it as political bias is because the description is less than flattering.

The chapter on the Byzantine Empire, the "Lands of the Romans", gives a valuable first-hand observation. It provides an observation-based account of the 360 languages spoken in the Caucasus, with the Lingua Franca being Arabic and Persian across the region.

With the description of Kiev, he describes the route of the Volga Bulgars and the Khazars.

He also published a cartographic map of Sindh together with accounts of the geography and culture of Sindh Valley of the Indus River.

Biographical Summary

Details known of Ibn Hawqal's life are extrapolated from his book. He spent the last 30 years of his life traveling to remote parts of Asia and Africa and writing about what he saw. One journey brought him 20° south of the equator along the East African coast where he discovered large populations in regions the ancient Greek writers had deemed were uninhabitable. The Greek had not spoken with knowledge; they were merely speculating based on some logic that they believed in.

The date of his death, known from his writings, was after 978 AD.

28. Ibn al-Qutiyya

Muḥammad Ibn ʿUmar Ibn ʿAbd al-ʿAzīz ibn ʾIbrāhīm ibn ʿIsā ibn Muzāḥim

(Arabic: محمد ابن عمر ابن عبد العزيى ابن إبراهيم ابن عيسى ابن مزاحم),

also known as Ibn al-Qūṭiyya (ابن القوطية), Abu Bakr, or al-Qurtubi ("the Córdoban"),

(died 6 November 977),

was a historian and a philologist. His magnum opus, the History of the Conquest of al-Andalus, is one of the earliest accounts of the Islamic conquest of Spain.

Contributions in Social Sciences

Ta'rikh iftitāḥ al-Andalus (تاريخ افتتاح الأندلس), 'History of the Conquest of al-Andalus' is found in only a single extant manuscript, Bibliothèque Nationale de France No. 1867. Speculation about a copy's existence among the rich manuscript collection at Constantine, Algeria, of Si Hamouda ben Cheikh el-Fakoun, seems unlikely. The 18-volume history was written at the height of the Umayyad Caliphate of al-Andalus. It covers the first 250 years of the Umayyad Caliphate. Ibn al-Quṭīyya treats the lives of Christians, Jews and Muslim converts; and in addition to accounts of the rulers, it covers the intrigues among servants, minor officials, poets, judges, concubines and physicians.

There are following additional works of al-Quṭīyya.

Kitāb Taṣārīf al-afʿāl, ('Book on the Conjugation of Verbs')—The oldest MS of an Arabic dictionary extant.

Kitāb al-Maqṣūr wa 'l-Mamdūd ('Book on the Shortened and Extended Alif'); this title is mentioned by al-Faraḍī but no copy survives.

There are following Editions and translations:

- *Ta'rīj iftitāḥ al-Andalus*, critical transcript of the unique manuscript edited by P. de Gayangos (with collaboration by E. Saavedra and F. Codera), 1868.
- Historia de la conquista de España de Aben al-Cotia el cordobés, seguida de fragmentos históricos de Abencotaiba (y la noble carta dirigida a las comarcas españolas del wazīr al-Gassānī), Spanish translation by Julián Ribera, Madrid, 1926.
- *Early Islamic Spain: the History of Ibn al-Qūṭīya*, English translation by David James, Routledge, 2009.

Biographical Summary

He was from Andalusia, and served at the Umayyad court of caliph Al-Hakam II.

Ibn al-Qūṭiyya, whose name means "son of the Gothic woman", claimed descent from Wittiza, the last king of the united Visigoths in Spain, through a granddaughter, Sara the Goth, who travelled to Damascus and married ʿĪsā ibn Muzāḥim. Sara and ʿĪsā then returned to Al-Andalus.

Ibn al-Qūṭiyya was born and raised in Seville. His family was under the patronage of the Qurayshi tribe, and his father was a qāḍī (judge) in Seville and Écija. The Banu Hajjaj, also of Seville, were close relatives of his family, also claiming descent from Visigothic royalty.

Ibn al-Qūṭiyya's student al-Faraḍī composed a short sketch of his master for his biographical dictionary, preserved in a late medieval manuscript discovered in Tunis in 1887.

Ibn al-Qūṭiyya died in 977 AD.

29. Al-Maqdisi

Shams al-Dīn Abū ʿAbd Allāh Muḥammad ibn Aḥmad ibn Abī Bakr al-Maqdisī

(Arabic: شَمْس ٱلدِّيْن أَبُو عَبْد ٱلله مُحَمَّد ابْن أَحْمَد ابْن أَبِي بَكْر ٱلْمَقْدِسِي),

also known as al-Maqdisī (Arabic: ٱلْمَقْدِسِي),

(c. 945/946 – 991),

was father of geography, author of Aḥsan al-taqāsīm fī maʿrifat al-aqālīm (The Best Divisions in the Knowledge of the Regions), as well as author of the book, Description of Syria (Including Palestine).

Contributions in Social Sciences

Al-Maqdisi excelled his predecessors in being the first to desire and conceive geography as an "original science". This is an assertion that al-Maqdisi himself makes in the preface of Aḥsan al-taqāsīm.

He belonged to the school known as the "atlas of Islam", inaugurated by Abu Zayd al-Balkhi (d. 934) and developed by Istakhri (d. 957) and al-Maqdisi's contemporary Ibn Hawqal (d. 978).

Al-Maqdisi refers to this world as al-mamlaka al-Islām (the Domain of Islam), which is a first as a unique concept in which all of the lands of Islam constituted a single domain. He subdivided this domain into two parts: mamlakat al-ʿArab (domain of the Arabs) and mamlakat al-ʿAjam (domain of the non-Arabs). The former consisted, from east to west, of the six provinces of Iraq, Aqur (Upper Mesopotamia), Arabia, Syria, Egypt and the Maghreb, while the latter consisted of the eight provinces of the Mashriq (Sistan, Afghanistan,

Khurasan and Transoxiana), Sindh, Kirman, Fars, Khuzistan, Jibal, Daylam and Rihab (Armenia, Adharbayjan and Aran).

Aḥsan al-taqāsīm gives a systematic account of all the places and regions al-Maqdisi had visited. He devoted a section of his book to Bilad al-Sham (the Levant) with a particular focus on Palestine. In contrast to travelers to Palestine, such as Arculf (c. 680s), Nasir Khusraw (c. 1040s) and others, who were pilgrims, al-Maqdisi gave detailed insights into the region's population, way of life, economy and climate. He paid special attention to Jerusalem, detailing its layout, walls, streets, markets, public structures and landmarks, particularly the Haram ash-Sharif (Temple Mount) and the latter's Dome of the Rock and al-Aqsa Mosque. He described the city's people and customs, focusing on its Muslims, but also its Christian and Jewish communities.

Al-Maqdisi also gave extensive overviews of Ramla and Tiberias, the capitals of the Palestine and Jordan districts, respectively. To a lesser extent, he described Acre, Beisan, Bayt Jibrin, Caesarea, Amman and Aila. In his descriptions of the aforementioned cities, al-Maqdisi noted their prosperity and stability and gave a general impression of Palestine as densely populated and wealthy, with numerous localities.

His description of Palestine, and especially of Jerusalem, his native city, all that he wrote, is the fruit of his own observation, and his descriptions of the manners and customs of the various countries, bear the stamp of a scientifically observant mind, fortified by profound knowledge of both books and men.

In describing the Eastern Arabia, Al-Maqdisi says in *985* CE that:

Hafit [Tuwwam] abounds in palm trees. It lies in the direction of Hajar [Al-Hasa], and the mosque is in the place. Dibba and Julfar, both in the direction of the Hajar, are close to the sea; and Tuwwam has been dominated by a branch of the Quraysh.

Al-Maqdisi mentioned regions in Eastern Arabia which form parts of what are now Saudi Arabia, the UAE and Oman. Al-Hasa is an important oasis region in the eastern part of Saudi Arabia, whereas Tuwwam is another oasis region split between the UAE and Oman, comprising the modern settlements of Al Ain and Al-Buraimi on different sides of the Omani-UAE border. Dibba is another region split between the UAE and Oman, touching the Musandam Peninsula, which is partly ruled by the Emirate of Ras Al Khaimah, where the ancient settlement of Julfar is located.

Biographical Summary

He is one of the earliest known historical figures to self-identify as a Palestinian during his travels.

Al-Maqdisi was born in Jerusalem in ca. 946 and belonged to a middle class family whose roots in the city's environs dated from the period approximate to the 7th-century Muslim conquest. He was very much attached to the Palestine of his birth and to the town whose name he bears. Al-Muqaddasī was a nisba indicating that he was from "Bayt al-Muqaddas", the Arabic name for Jerusalem. His paternal grandfather, Abu Bakr al-Banna, had been responsible for the construction of Acre's maritime fortifications under orders from Ahmad ibn Tulun (r. 868–884), the autonomous Abbasid governor of Egypt and Syria. Al-Maqdisi's maternal grandfather, Abu Tayyib al-Shawwa, moved to Jerusalem from Biyar in Khurasan and was an architect.

Al-Maqdisi was well-educated; his use of "rhymed prose, even poetry" reflects a strong knowledge in Arabic grammar and literature.

His writings show that he possessed an interest in Islamic jurisprudence, history, philology and hadith.

Al-Maqdisi made his first Hajj (pilgrimage to Mecca) in 967. During this period, he became determined to devote himself to the study of geography. To acquire the necessary information, he undertook a series of journeys throughout the Islamic world, ultimately visiting all of its lands with the exception of al-Andalus (Iberian Peninsula), Sindh and Sistan. The known dates or date ranges of al-Maqdisi's travels include his journey to Aleppo sometime between 965 and 974, his second pilgrimage to Mecca in 978, a visit to Khurasan in 984 and his stay in Shiraz in 985 when he decided to compose his material. The finished work was titled Aḥsan al-taqāsīm fi maʾarfat al-aqalīm (The Best Divisions for the Knowledge of the Provinces).

Al-Maqdisi died in 991 AD.

30. Ibn al-Nadim

Abū al-Faraj Muḥammad ibn Isḥāq al-Nadīm

(Arabic: ابو الفرج محمد بن إسحاق النديم),

also ibn Abī Ya'qūb Isḥāq ibn Muḥammad ibn Isḥāq al-Warrāq, and commonly known by the nasab (patronymic) Ibn al-Nadīm (Arabic: ابن النديم),

(died 17 September 995 or 998),

was a bibliographer and biographer from Baghdad who compiled the encyclopedia Kitāb al-Fihrist (The Book Catalogue).

Contributions in Social Sciences

The Kitāb al-Fihrist (Arabic: كتاب الفهرست) is a compendium of the knowledge and literature of tenth-century Islam, including approximately 10,000 books and 2,000 authors. This crucial source of literature preserves from his own hand the names of authors, books and accounts otherwise entirely lost. Al-Fihrist is evidence of Al-Nadim's thirst for knowledge among the exciting sophisticated milieu of Baghdad's intellectual elite. As a record of civilization, it provides unique classical material.

Biographical Summary

Much known of Ibn al-Nadīm is deduced from his epithets. 'Al-Nadim' (النَّدِيم), 'the Court Companion' and 'al-Warrāq (الْوَرَّاق) 'the copyist of manuscripts'.

Probably born in Baghdad ca. 320/932 he died there on Wednesday, 20th of Shaʿban A.H. 385.

From age six, he may have attended a madrasa and received comprehensive education in Islamic studies, history, geography, comparative religion, the sciences, grammar, rhetoric and Qur'anic commentary. Ibrahim al-Abyari, author of Turāth al-Insaniyah says al-Nadim studied with al-Hasan ibn Sawwar, a logician and translator of science books; Yunus al-Qass, translator of classical mathematical texts; and Abu al-Hasan Muhammad ibn Yusuf al-Naqit.

An inscription, in an early copy of al-Fihrist, probably by the historian al-Maqrizi, relates that al-Nadim was a pupil of the jurist Abu Sa'id al-Sirafi (d.978/9), the poet Abu al-Faraj al-Isfahani, and the historian Abu Abdullah al-Marzubani and others. Al-Maqrizi's phrase 'but no one quoted him', would imply al-Nadim himself did not teach.

While attending lectures of some of the leading scholars of the tenth century, he served as an apprentice in his father's profession, the book trade.

His father was a bookdealer and owner of a prosperous bookstore. He commissioned al-Nadim to buy manuscripts from dealers. Al-Nadim, with the other calligrapher scribes employed, would then copy these for the customers.

The bookshop, customarily on an upper floor, would have been a popular hangout for intellectuals.

Ibn al-Nadīm probably visited the intellectual centers at Basra and Kufa in search of scholarly material. He may have visited Aleppo, a center of literature and culture under the rule of Sayf al-Dawla.

Ibn al-Nadīm may have served as 'Court Companion' to Nasir al-Dawla, a Hamdanid ruler of Mosul who promoted learning. His family were highly educated and he, or his ancestor, may have been a 'member of the Round Table of the prince'.

The physician Ibn Abi Usaibia (d. 1273), mentions al-Nadim thirteen times and calls him a writer, or perhaps a government secretary.

In 987, Ibn al-Nadim began compiling al-Fihrist (The Catalogue), as a useful reference index for customers and traders of books. Over a long period, he noted thousands of authors, their biographical data, and works, gathered from his regular visits to private book collectors and libraries across the region - including Mosul and Damascus - and through active participation in the lively literary scene of Baghdad.

31. Al-Musabbihi

Al-Amīr al-Mukhtār ʿIzz al-Mulk Abū ʿAbd Allāh Muḥammad ibn Abi'l Qāsim ʿUbayd Allāh ibn Aḥmad ibn Ismāʿīl ibn ʿAbd al-Azīz al-Ḥarranī al-Musabbiḥī al-Kātib, commonly known simply as al-Musabbihi

(Arabic: المصبحي),

(4 March 977 – April/May 1030),

was a Fatimid historian, writer and administrative official.

Contributions in Social Sciences

Al-Musabbihi is known to have authored some 40,000 pages of manuscripts dealing with an array of topics, including history, psychology, law, grammar, sexology and cooking.

Akhbār Miṣr, a chronicle of Egyptian history and news, was among al-Musabbihi's well-known works. However, like the vast majority of al-Musabbihi's works, only fragments of Akhbār Miṣr survived; most of his writings disappeared not long after his death.

Al-Musabbihi was a prolific writer, who authored numerous manuscripts on a variety of subjects, including history, practical psychology, sexology, law, grammar and cooking. The later medieval historians Ibn Khallikan and Ibn Sa'id al-Andalusi documented in detailed lists all of al-Musabbihi's works and the number of pages for each work. Al-Musabbihi's total work amounted to roughly 40,000 pages, with some works alone consisting of over a thousand pages. However, most of his work disappeared not long after he died.

Among al-Musabbihi's main works was the roughly 13,000-page chronicle of Egypt's history, known as Akhbār Miṣr. Like al-Musabbihi's other works, much of Akhbār Miṣr was lost early on, with exceptions including quotations of this work that were copied in other historians' compilations.

The only known manuscript of Akhbār Miṣr that was preserved is chapter 40, which is located in the Escorial in Spain. This section of al-Musabbihi's work documented events occurring in the caliphate between 1023/24 and 1024/25 and also contained poems and letters by other writers who were well-acquainted with al-Musabbihi. This manuscript provides an insight into the contemporary literary composition, in elegant prose and poetry, of Egypt and Iraq at the beginning of the 11th century. The modern Egyptian historian Ayman Fuad Sayyid claims the 15th-century Egyptian historian al-Maqrizi had in his possession chapter 34 of Akhbar Misr, which recorded events for the year 1004/05.

Akhbar Misr was a contemporary work in which al-Musabbihi recorded the day-to-day events in the Fatimid Caliphate and at the end of the year, recorded the obituaries of notable individuals.

As an administrative official in Cairo, he also documented the often-violent struggle for power in the aftermath of al-Hakim's death, centered around various military commanders and civil officials, who utilized informers and forged documents to discredit each other in front of al-Hakim's child successor, az-Zahir (r. 1021–1036). He held suspicions about the intrigues of Sitt al-Mulk's entourage with az-Zahir. Al-Musabbihi also included news about the Bedouin uprising against the Fatimids in Syria in 1024/25 that placed the Mirdasids and Jarrahids in power there. Moreover, Akhbār Miṣr recorded regular aspects of life in Fustat, ranging from road accidents and crimes to wholesale and retail prices on goods amid a famine to hippopotami roaming in the Nile River.

Biographical Summary

Al-Musabbihi was born in Fustat, Egypt, on 4 March 977. His family was originally from Harran in the Jazira (Upper Mesopotamia). Not common for a locally-born Sunni civilian, al-Musabbihi joined the Fatimid military. He was made governor of al-Qays and Bahnasa (both in Upper Egypt) and held the title of amir (commander). He was later appointed head of the dīwān al-tartīb, a position equivalent to general secretary of the central administration. He traveled daily from al-Fustat to his government post in Cairo and on most evenings, he stopped by the historic Mosque of Amr ibn al-As and interacted with his and his father's friends, most of whom were Syrian Muslim traditionalists. Though he was a devout Sunni Muslim, al-Musabbihi was loyal to the Fatimids' Ismaili Shia state and maintained a particularly close relationship with the eccentric caliph al-Hakim (r. 996–1021). The latter was known to get along with the residents of Fustat.

Al-Musabbihi was a devout Sunni Muslim born in Fustat, where he lived most of his life and died. He was known to be loyal to the Fatimid government and maintained particularly close ties with Caliph al-Hakim (r. 996–1021). Early in his career, he served in the Fatimid military and was made a provincial governor in Upper Egypt before becoming a leading figure in the Fatimids' central administration in Cairo.

Al-Musabbihi died in Fustat in April/May 1030.

32. Abu al-Wafa' Al-Mubashshir ibn Fatik

Abu al-Wafa' al-Mubashshir ibn Fatik

(Arabic: ابو الوفاء المبشّر بن فاتك),

was a polymath, versed in the mathematical sciences and also wrote on logic and medicine. He authored Kitāb mukhtār al-ḥikam wa-maḥāsin al-kalim (مختار الحكم ومحاسن الكلم), the "Book of Selected Maxims and Aphorisms".

Contributions in Social Sciences

Al-Mubaššir ibn Fātik authored Kitāb mukhtār al-ḥikam wa-maḥāsin al-kalim (مختار الحكم ومحاسن الكلم), the "Book of Selected Maxims and Aphorisms". It was a collection of biographies of twenty-one "sages". His al-Mukhtar was a great success in the centuries that followed. It provided source material for later scholars; such as Muhammad al-Shahrastani in his book Kitab al-wa-l-Milal Nihal and Shams al-Din al-Shahrazuri for his Nuzhat al-Arwah.

The book was translated in European languages. Through these translations the book was a valuable source for European writers. Following are some of the translations.

- Los Bocados de Oro; translated in the reign of Alfonso X of Castile (1252–1284) was the earliest translation into a Western European vernacular (Spanish).
- Liber Philosophorum Moralium Antiquorum by the Italian John of Procida (1298), friend and doctor of Emperor Frederick

II. Several early Latin translations appeared as florilegia and excerpts integrated into larger works (Latin).

- Les Dits Moraulx des Philosophes by Guillaume de Tignonville [fr], chamberlain to King Charles VI; Middle French from the Latin translation. Of the fifty manuscripts extant the oldest dates from 1402. The first printed editions were made in Bruges by Colard Mansion (no date, perhaps 1477), in Paris by Antoine Vérard in 1486, by Jean Trepperel in 1502, by Galliot du Pré in 1531, etc. (Nine reported editions by 1533) (French).
- Los Dichs dels Philosophes from the Tignonville's French translation (Occitan).
- The Dicts or Sayings of the Philosophers (1450) by Stephen Scrope for his stepfather, John Fastolf; Middle English translation (English).
- The Dictes or Sayengis of the Philosophhres (1473) by Anthony Woodville. William Worcester amended Woodville's translation and it appears this was the version printed by William Caxton in his Westminster workshop on November 18, 1477, the first book printed in England, that is discounting Thomas Aquinas's Apostles' Creed, (Expositio in Symbolum Apostolorum) printed December 17, 1468 (English).

In addition, there were multiple editions of the book.

Arabic

- ʿAbd al-Raḥmān Badawī (ed.), *Mukhtār al-ḥikam wa-maḥāsin al-kalim*, Publicaciones del Instituto de Estudios Egipcio

Islámicos (Egyptian Institute for Islamic Studies), Madrid, 1958.

- Before this edition, only the Lives of Alexander the Great and Aristotle had been published:
- Bruno Meissner, "Mubachchir's Akhbar al-Iskandar". Zeitschrift der Deutschen Gesellschaft morgenländischen, vol. 49, 1895, pp. 583–627.
- Julius Lippert, Studien auf dem Gebiete der griechisch-arabischen Übersetzungsliteratur. Heft I, Brunswick, Richard Sattler, 1894, pp. 3–38 ("Quellenforschungen zu den arabischen Aristoteles-biographien").

Medieval Spanish

- Hermann Knust (ed.), "Este libro es llamado bocados de oro, el qual conpuso el rrey Bonium, rrey de Persia". Mittheilungen aus dem Eskurial, Bibliothek des literarischen Vereins in Stuttgart, CXLIV, Tübingen, 1879, pp. 66–394.
- Mechthild Crombach (ed.), "Bocados de oro: Seritische Ausgabe des altspanischen Textes". Romanistische Versuche und Vorarbeiten, 37. Romanischen Seminar der Universität Bonn, Bonn, 1971.

Latin

- Ezio Franceschini (ed.), "Liber philosophorum moralium antiquorum". Atti del Reale Istituto Veneto di Scienze, Lettere ed Arti, vol. 91, No. 2, 1931–1932, pp. 393–597.

Medieval French

- Robert Eder (ed.), "Tignonvillana inedita". Romanische Forschungen, vol. 33. (Ludwig-Maximilians-Universität München), Erlangen, Fr. Junge, 1915, pp. 851–1022.

Middle English

- William Blades, The Dictes and Sayings of the Philosophers. A facsimile reproduction of the first book printed in England by William Caxton in 1477, (translated from the Medieval French by Anthony, Earl Rivers; edited by William Caxton). London: Elliot Stock, 1877. Although three subsequent editions of the book were printed in Caxton's lifetime, of the first of these editions, the only surviving copy carrying Caxton's printer's mark and dated November 18, 1477, is held at the John Rylands Library, Manchester.

Biographical Summary

He was born in Damascus but lived mainly in Egypt during the 11th century Fatimid Caliphate.

Ibn Abi Usaibia wrote Uyūn ul-Anbāʾ fī Ṭabaqāt ul-Aṭibbā (عيونالأنباء في طبقات الأطباء), "the History of Physicians". According to this book, Ibn Fatik was from a noble family and held the position of "emir" at the court of the Fatimids in the reign of al-Mustansir Billah. He was a passionate bibliophile, acquired a great collection of books and enjoyed the company of scholars. He devoted himself to study and research. He trained in mathematics and astronomy under the philosopher, mathematician and astronomer Ibn al-Haytham. He also associated with Ibn al-Amidi and the physician, astrologer, and astronomer Ali ibn Ridwan (988–1061).

When Al-Mubaššir ibn Fātik died, many heads of state attended his funeral. According to this biography, such was his wife's disaffection through want of attention, she threw most of his books into the pool at the center of the house, and so they were lost.

33. Al-Maʿarri

Abū al-ʿAlāʾ Aḥmad ibn ʿAbd Allāh ibn Sulaymān al-Tanūkhī al-Maʿarrī

(Arabic: أبو العلاء أحمد بن عبد الله بن سليمان التنوخي المعري),

(December 973 – May 1057),

was a philosopher, poet, and writer.

Contributions in Social Sciences

Following are the literary works of al-Maʿarrī.

- An early collection of his poems is Saqṭ al-zand; سقط الزند) meaning The Tinder Spark. It gained great popularity and established his reputation as a poet.
- A second collection appeared under the title Luzūm mā lam yalzam لزوم ما لا يلزم, or simply Luzūmīyāt اللزوميات. The title refers to how al-Ma'arri saw the business of living and alludes to the unnecessary complexity.
- His third famous work is in prose known as Resalat Al-Ghufran رسالة الغفران. The work was written as a response to the Arabic poet Ibn al-Qarih. In this work, the poet visits paradise and meets the Arab poets of the pagan period.
- Dante wrote his Divine Comedy centuries later, and may have gotten his clues from Resalat Al-Ghufran. However, European writers generally feel no obligation to maintain integrity by acknowledging their references in Muslim civilization.

- Algeria banned Resalat Al-Ghufran from the International Book Fair held in Algiers in 2007.
- Al-Fuṣūl wa al-ghāyāt is a collection of moral lessons (العظة).
- A corpus of verse riddles.

Following are modern editions of Saqṭ al-zand; سقط الزند.

- Risalat ul Ghufran, a Divine Comedy. Translated by G. Brackenbury 1943.
- The Epistle of Forgiveness: Volume One: A Vision of Heaven and Hell. Translated by Geert Jan Van Gelder and Gregor Schoeler. Library of Arabic Literature, New York University Press 2013.
- The Epistle of Forgiveness: Volume Two: Hypocrites, Heretics, and Other Sinners. Translated by Geert Jan Van Gelder and Gregor Schoeler. Library of Arabic Literature, New York University Press 2014.
- Those riddles of al-Maʿarrī that are cited in al-Ḥaẓīrī's twelfth-century Kitāb al-Iʿjāz fī l-aḥājī wa-l-alghāz have been edited as Abū l-ʿAlāʾ al-Maʿarrī, Dīwān al-alġāz, riwāyat Abī l-Maʿālī al-Ḥaẓīrī, ed. by Maḥmūd ʿAbdarraḥīm Ṣāliḥ (Riyadh [1990]).

Biographical Summary

Abu al-'Ala' was born in December 973 in al-Ma'arra (present-day Ma'arrat al-Nu'man, Syria), southwest of Aleppo. He was a member of the Banu Sulayman, a notable family of Ma'arra, belonging to the larger Tanukh tribe. One of his ancestors was probably the first qadi of Ma'arra. The Tanukh tribe had formed part of the aristocracy in Syria for hundreds of years and some members of the Banu Sulayman had also been noted as good poets.

He lost his eyesight at the age of four due to smallpox.

He started his career as a poet at an early age, at about 11 or 12 years old. He was educated at first in Ma'arra and Aleppo, later also in Antioch and other Syrian cities. Among his teachers in Aleppo were companions from the circle of Ibn Khalawayh. This grammarian and Islamic scholar had died in 980 CE, when al-Ma'arri was still a child. Al-Ma'arri nevertheless laments the loss of Ibn Khalawayh in strong terms in a poem of his Risālat al-ghufrān.

In 1004–05 al-Ma'arri learned that his father had died and, he wrote a Rithā' where he praised his father. Years later he would travel to Baghdad where he became well received in the literary salons of the time, though he was a controversial figure. After the eighteen months in Baghdad, al-Ma'arri returned home. He may have returned because his mother was ill. He returned home in about 1010 and learned that his mother had died before his arrival.

He remained in Ma'arra for the rest of his life, where he opted for an ascetic lifestyle, refusing to sell his poems. His personal confinement to his house was only broken one time when violence had struck his town. In that incident, al-Ma'arri went to Aleppo to intercede with its Mirdasid emir, Salih ibn Mirdas, to release his brother Abu'l-Majd and several other Muslim notables from Ma'arra who were held responsible for destroying a winehouse whose Christian owner was accused of a molesting a Muslim woman. Though he was confined, he lived out his later years continuing his work and collaborating with others. He enjoyed great respect and attracted many students locally, as well as actively holding correspondence with scholars. Despite his intentions of living a secluded lifestyle, in his seventies, he became rich and was the most revered person in his area. Al-Ma'arri never married and died in May 1057 in his home town.

Al-Ma'arri was an ascetic, renouncing worldly desires and living secluded from others while producing his works. In Baghdad, while being well received, he decided not to sell his texts, which made it difficult for him to live.

34. Al-Bakri

Abū ʿUbayd ʿAbd Allāh ibn ʿAbd al-ʿAzīz ibn Muḥammad ibn Ayyūb ibn ʿAmr al-Bakrī

(Arabic: أبو عبيد عبد الله بن عبد العزيز بن محمد بن أيوب بن عمرو البكري),

or simply al-Bakrī,

(c. 1040 – 1094),

was a historian and a geographer from Andalusia.

Contributions in Social Sciences

Al-Bakri wrote about Europe, North Africa, and the Arabian peninsula. Only two of his works have survived. His Mu'jam mā ista'jam contains a list of place names mostly within the Arabian peninsula with an introduction giving the geographical background. His most important work is his Kitāb al-Masālik wa-al-Mamālik ("Book of Highways and of Kingdoms") (كتاب المساليك و المماليك). This was composed in 1068, based on literature and the reports of merchants and travelers. It is one of the most important sources for the history of West Africa and gives crucial information on the Ghana Empire, the Almoravid dynasty and the trans Saharan trade. He also included information on events that occurred close to the time that he wrote.

Al-Bakri mentions the earliest urban centers in the trans Saharan trade to embrace Islam, late in the 10th century, Gao was one of the very few along the Niger River to have native Muslim inhabitants. Other centers along the serpentine bends of the great river eventually followed: Takrur (Senegal); Songhay (Mali); Kanem-Bornu (Chad);

and Hausa-territories (Nigeria). By the 11th century, reports on these and other flourishing Islamic cities made their way north to Al-Andalus in southern Iberia, enabling Al-Bakri to write in his Kitab al-Masalik wa al-Mamalik (Book of Highways and Kingdoms): "The city of Ghana consists of two towns situated on a plain", "One of these towns, which is inhabited by Muslims, is large and possesses twelve mosques in one of which they assemble for the Friday prayer. There are salaried Imams and Muezzins, as well as Jurists and Scholars."

His works are noted for the relative objectiveness with which they present information. For each area, he describes the people, their customs, as well as the geography, climate and main cities. Similar information was also contained in his written geography of the Arabian Peninsula, and in the encyclopedia of the world in which he wrote. He also presented various anecdotes about each area. Unfortunately, parts of his main work have been lost, and of the surviving parts, some have never been published.

The crater Al-Bakri on the Moon is named after him.

Biographical Summary

Al-Bakri was born in 1040 AD in Huelva. He was the son of the sovereign of the short-lived principality there. He belonged to the Arab tribe of Bakr. When his father was deposed by al-Mu'tadid he moved to Córdoba, where he studied with the geographer al-Udri and the historian Ibn Hayyan. He spent his entire life in Al-Andalus, most of it in Seville and Almeria. He died in Córdoba in 1094 AD, without ever having travelled to the locations of which he wrote.

35. Al-Humaydi al-Azdi

Abu Abd Allah Muhammad ibn Abi Nasr Futuh ibn Abd Allah ibn Futuh ibn Humayd ibn Yasil, most commonly known as al-Humaydi Al-saboni,

was a scholar of history and Islamic studies from Andalus.

Contributions in Social Sciences

Humaydi was famous for his biography of the notables of Islamic Spain, entitled Jadhwat al-muqtabis fī tārīkh ʻulamā' al-Andalus (جذوة المقتبس فى ذكر ولاة الاندلس) OCLC 13643176. He composed the book while in Baghdad on request of his friends, writing entirely from memory without any other written sources. The book is considered the earliest primary source to mention Abu al-Qasim al-Zahrawi, and an important primary source for the life of Ziryab.

Humaydi's historical works were the main primary source on the Pisan–Genoese expeditions to Sardinia in the early 11th century, largely considered precursors to the Crusades.

In the field of hadith, Humaydi is credited with inventing the genre of combining multiple independent books of hadith into bound collections, a style of cataloging which would gain even more popularity in the 12th century. His books on hadith are also considered significant to modern attempts at critical reevaluations, especially al-Jamʻ bayna al-Ṣaḥīḥayn (الجمع بين الصحيحين) OCLC 41454057, which is his linguistic commentary on the two most important canonical works, Sahih al-Bukhari and Sahih al-Muslim.

Following are some of his Edited works:

- al-Dhahab al-masbuk fi wa'z al-muluk. Eds. Abu Abd al-Rahman Ibn Aqil al-Zahiri and Dr. Abd al-Halim Uways. Riyadh: Dar Alam al-Kutub, 1982. 235 pages. (Kings and rulers).
- El Tafsir al-gharib ma fi al-Sahihayn de el Humaydi. Ed. Z.M.S. 'Abd al-'Aziz (PhD dissertation). Complutense University of Madrid, 1989. Based on a 1995 reprint of Humaydi's original from Maktabat al-Sunna in Cairo.

Biographical Summary

Humaydi's family belonged to the Arab Azd tribe from Yemen. According to the Encyclopedia of Islam, his father was born in al-Rusafa, a suburb of Córdoba. Due to civil strife at the time, Humaydi's father moved to the island of Majorca, where Humaydi was born in c. 1029AD.

While in Spain, Humaydi was a student of Ibn 'Abd al-Barr and both a student and friend of Ibn Hazm, from whom Humaydi took his Zahirite views in Muslim jurisprudence. Due to persecution of Zahirites in Al-Andalus by the rival Malikites at the time, Humaydi fled from Spain for good in 1056. Initially, he went to Mecca and performed the Muslim pilgrimage before traveling to Tunisia, Egypt and Damascus to pursue Hadith studies. Like many scholars of that field, Humaydi frequently worked with manuscripts written in different eras. He was an outstanding scholar in the fields of history, Arabic grammar as well as lexicography.

Eventually, Humaydi settled down in Baghdad, where the Zahirite had once been the official law of the land. While not enjoying state sponsorship, his views did receive tolerance as opposed to the outright persecution from which Humaydi had escaped. He died in the city in 1095.

36. Ibn Bassam

Ibn Bassām al-Shantarinī

(Arabic: ابن بسام الشنتريني),

(1058-1147),

was a poet and historian from al-Andalus.

Contributions in Social Sciences

Ibn Bassam describes how the incessant invasions of the Castillans forced him to run away from Santarém in Portugal, "the last of the cities of the west," after seeing his lands ravaged and his wealth destroyed, a ruined man with no possessions, save his battered sword.

Especially well known is his anthology Dhakhīra fī mahāsin ahl al-Jazīra (The Collection concerning the Merits of the People of Iberia). It is one of the most important sources of information in the field of history, literature and culture of the Almoravid dynasty. It was edited in eight volumes by Ihsan Abbas, written in rhymed prose.

Many of its biographies are contemporary and filled out with details taken from the Kitab al-Matin of Ibn Hayyan. The parts taken from that book are easily distinguishable, because Ibn Bassam prefixes the words qala Ibn Hayyan ("Ibn Hayyan says") and concludes the extract with intaha kalam Ibn Hayyan ("here ends lbn Hayyan's words").

Following are some Editions and translations of Dhakhīra fī mahāsin ahl al-Jazīra:

- 'Abī 'al-Ḥasan 'Alī ibn Bassām 'al-Shantarīnī, 'al-Dhakhīrah fī maḥāsin ahl 'al-Jazīrah, ed. by Iḥsān 'Abbās, 4 vols in 8

(Bayrūt: Dār 'al-Thaqāfah, 1978-81), https://al-maktaba.org/book/1035, https://archive.org/details/zakhera_mahasen_jazeera

- 'Ibn Bassām, from Al-dhakhīra fī maḥāsin ahl al-Jazīra translation', trans. by Ross Brann, in Medieval Iberia, ed. by Remie Constable, 2nd edn (Philadelphia: University of Pennsylvania Press, 2011), pp. 125-27.

Biographical Summary

Ibn Bassām was born in 1058 AD in Santarém (sometimes spelled Shantarin). He was a poet and historian from al-Andalus. He hailed from the Banu Taghlib tribe. He died in 1147.

37. Ibn al-Qalanisi

Abu Ya'la Hamzah ibn Asad ibn al-Qalanisi
(Arabic: ابو يعلى حمزة ابن الاسد ابن القلانسي),
(c. 1071 – 18 March 1160),
was a historian in 12th-century Damascus.

Contributions in Social Sciences

Ibn al-Qalanisi wrote the Dhail or Mudhayyal Ta'rikh Dimashq ('Continuation of the Chronicle of Damascus'). It was an extension of the chronicle of Hilal bin al-Muhassin al-Sabi'; extending the chronicle till al-Qalanisi's death in 1160.

This chronicle is one of the few contemporary accounts of the First Crusade and its immediate aftermath, making it not only a valuable source for modern historians, but also for later 12th-century chronicles, including Ali ibn al-Athir. He also witnessed the siege of Damascus in 1148 during the Second Crusade, which ended in a decisive crusader defeat.

The entire material of his chronicle covers the time span of two generations, his father's and his own. Al-Qalanisi experienced the First Crusade at a mature age, although apparently not as a fighter.

In describing his modus operandi, Al-Qalanisi says that he extracted his information both from eyewitnesses, including himself, and documents. As a result of al-Qalanisi's careful work, the chronology of events is very accurate. He even offers the day of the week when the event took place.

Gibb extracted from the chronicle the period 1097–1159, and translated it to English and published it in 1932.

Biographical Summary

Al-Qalanisi was born in 1071 AD, a descendent from the Banu Tamim tribe. He was among the well-educated nobility of Damascus. He studied literature, theology, and law. He served firstly as a secretary and later as the head of the chancery of Damascus (the Diwan al-Rasa'il). He served twice as ra'is of the city, an office equivalent to mayor.

Al-Qalanisi died in 1160.

38. Abu'l-Hasan al-Bayhaqi

Zahir al-Din Abu'l-Hasan Ali ibn Zayd-i Bayhaqi

(Persian: ظهیرالدین ابوالحسن علی بن زید بیهقی),

also known as Ibn Fondoq (ابن فندق),

(c. 1097 – 1169),

was an Iranian polymath and historian of Arab descent.

Contributions in Social Sciences

Bayhaqi is the author of Tarikh-i Bayhaqi.

Bayhaqi authored over 70 works in astronomy, history, Arabic grammar, philology.

Biographical Summary

Bayhaqi was a descendant of Khuzaima ibn Thabit (died 657), a companion of Prophet Muhammad. Most of his forefathers were either judges or Imams.

Bayhaqi was born in 1097 in Sabzevar, in northeastern Iran, the main city of the Bayhaq district, where his father's estates were located.

In 1114, Bahyaqi along with his father visited Omar Khayyam in Nishapur, who is a famous Persian mathematician and astronomer, as well as Philosopher who is famous for his Rubaiyat. While there, Bayhaqi began his education in literature and sciences. He moved to Marv where he completed his studies in Islamic jurisprudence by 1123. He returned to Nishapur in c.1127 when according to Yaqut al-Hamawi, he got married.

Bayhaqi became the qadi of Bayhaq through the efforts of his father-in-law, Shehab-al-Din Moḥammad b. Mas'ud, along with patronage from the Seljuq Sultan, Sanjar. He would soon resign that position.

Bayhaqi then traveled to Ray and devoted himself to mathematics and astronomy, and later further studied astronomy in Nishapur.

While living in Sarakhs he squandered all of his money, and returned to Nishapur.

After an attempt to establish himself in Bayhaq failed, Bayhaqi finally returned again to Nishapur and settled down to a life in the mosque and teaching and learning.

Bayhaqi died in 1169 AD.

39. Ibn Sidah

Abū'l-Ḥasan ʻAlī ibn Ismāʻīl

(Arabic: أبو الحسن على بن اسماعيل),

also known as Ibn Sīdah (ابن سيده), or Ibn Sīdah'l-Mursī (ابن سيده المرسي),

(c.1007-1066),

was a linguist, philologist and lexicographer of Arabic language. He was from Andalusia.

Contributions in Social Sciences

He compiled the encyclopedia al-Kitāb al-Mukhaṣṣaṣ (المخصص) (Book of Customs).

He researched and wrote the Arabic language dictionary Al-Muḥkam wa-al-muḥīt al-aʻẓam (المحكم والمحيط الأعظم) (The Great and Comprehensive Arbiter).

Biographical Summary

Ibn Sīdah was born in1007 AD in Murcia in eastern Andalusia. The historian Khalaf ibn ʻAbd al-Malik Ibn Bashkuwāl (ابن بشكوال) (1183-1101) in his book Kitāb aṣ-Ṣilah (كتاب الصلة) (Book of Relations) gives Ismāʻīl as the name of his father, in agreement with name given in the Mukhassas.

However, Al-Fath ibn Khaqan in mathmah al-anfus (مطمح الأنفس) has the name Aḥmad.

Yaqut al-Hamawi in The Lexicon of Literature, says about Ibn Sīdah that 'son of a woman' was his nickname.

Remarkably both he and his father were blind. His father was a sculptor although it seems the disciplines, he devoted his life to, philology and lexicography, had been in his family.

Mohammed ibn Ahmed ibn Uthman Al-Dhahabi's biographic encyclopedia Siyar A'lam al-Nubala (سير أعلام النبلاء) (Lives of The Noble Scholars) is the main biographic source. He lived in the taifa principality of "Dénia and the Eastern Islands" (طائفة دانية والجزائر الشرقية) under the rule of Emir Mujahid al-Amiri al-Muwaffaq (الأمير مجاهد العامري) (1044-1014).

He travelled to Mecca and Medina.

He studied in Cordova under the renowned grammarian Abu al-Sa'ad ibn al-Hasan al-Rubai al-Baghdadi (أبو العلاء صاعد بن الحسن الربعي البغدادي) (d.1026AD), who was exiled in Andalusia, and with Abu Omar al-Talmanki (أبي عمر الطلمنكي) (429-340AH).

He died in Dénia in 1066 AD.

40. Al-Jawaliqi

Abū Manṣūr Mauhūb al-Jawālīqī

(Arabic:_Arabic: أبو منصور الجواليقي),

(April 1074–17 July 1144),

was a grammarian of Arabic language, and a philologist.

Contributions in Social Sciences

Kitāb al-Mu'arrab (كتاب المُعَرَّب), (tr. 'Explanation of Foreign Words used in Arabic'), is his chief work. It was published as edited text, from an incomplete manuscript, by Eduard Sachau (Leipzig, 1867). Many of the lacunae in this have been supplied from another manuscript by W. Spitta in the Journal of the German Oriental Society, xxxiii. 208 sqq.

Al-Jawālīqī's Supplement to the Durrat ul-Ghawwas of Al-Hariri of Basra was published as Le Livre des locutions vicieuses; Arabic text with French introduction and notes was published by Hartwig Derenbourg, Morgenländische Forschungen (Leipzig, 1875), pp. 107–166.

Biographical Summary

Al-Jawālīqī was born in Baghdād in 1074, where he studied philology under Khātib al-Tibrizī (1030 - 1109).

He became famous for his hand writing.

He was a grammarian of Arabic language, and a philologist.

In his later years he acted as Imam to the Abbāsid caliph Al-Muqtafi.

He was born April 1074 and he died 17 July 1144.

41. Ibn Zafar al Siqilli

Hujjat al-Din Abu Abdallah Muhammad ibn Abi Muhammad ibn Muhammad ibn Zafar al-Siqilli

(Arabic: حجة الدين أبو عبد الله محمد بن أبي محمد بن محمد بن ظفر الصقلي),

commonly known simply as Ibn Zafar al-Siqilli,

was a polymath, philosopher and Arab-Sicilian politician of the Norman period (1104 - 1170).

Contributions in Social Sciences

Ibn Zafar authored 32 books.

Sulwān al-Muṭā fī Udwān al-Atbā (Arabic: سلوان المطاع في عدوان الأتباع) ('Consolation for the Ruler During the Hostility of his Subjects') is his magnum opus that was written during his stay in Sicily.

Michele Amari's Italian translation of this book appeared in 1851, seven centuries later. Amari's introduction had included a biographical account of Ibn Zafar and his manuscript's history.

Richard Bentley published an English translation in 1852.

At the beginning of the 20th century another Sicilian political scientist, Gaetano Mosca, wrote of the parallels between Ibn Ẓafar's treatise and Machiavelli's. Ibn Ẓafar is rarely credited as the precursor to Machiavelli. That is hardly surprising because the Western writers, as a rule, systematically refuse to give proper indebtedness to Muslim scientists.

The treatise is a form of wisdom literature with a long Arabian and Persian tradition, referred to as "mirrors for princes", which purported to be handbooks for princes and caliphs offering counsel on the proper use of power, good governance and the conduct of commerce and trade.

Ibn Zafar dedicated the first edition of Sulwan to an unknown king facing revolt - possibly the ruler of Damascus expelled by Nur ad-Din. He dedicated the second edition to his patron Abu'l-Qasim ibn Hammud ibn al-Hajar.

Ibn Zafar also authored a Biography of Illustrious Men. It has been translated into Italian, English and Turkish.

Biographical Summary

Ibn Zafar was said to be physically small and frail. His nisbah "al-Siqillī" indicates he was born in Sicily, but the patronym "al-Makkī" suggests his family origins were in Mecca, where he is believed to have been raised and educated. Nicknamed 'The Wanderer', the precise chronology of his travels is not known.

After a period in Sicily, Ibn Zafar first went to Fatimid Egypt, Mahdia in Tunisia, and then to Aleppo in 1146. In Aleppo, he taught at the Madrasa Ibn Abi Asrun under the patronage of Safi al-Din.

In 1154 he returned to Sicily under the patronage of Abu'l-Qasim ibn Hammud ibn al-Hajar, a Sicilian Arab noble. Due to the civil unrest of the Muslim population, sometime later, Ibn Zafar left Sicily definitively and took refuge in Hamat, in Syria, where he died in poverty in 1170, or 1172.

The geographer Yaqut al-Hamawi referred to him as a 'refined philologist', and both Shams al-Din al-Dhahabi and Ibn Khallikan praised his scholarship and thought.

42. Al-Suhayli

Sidi Abu al-Qasim Abd al-Rahman ibn Abd Allah al-Suhayli

(Arabic: أبو القاسم السهيلي),

(1114 – 1185),

was a historian of the Sirah of the Prophet, a grammarian, and a saint.

Contributions in Social Sciences

Al-Suhayli especially known as an Islamic scholar by his commentary on the sira of Ibn Hisham.

Al-Suhayli also wrote books on grammar and Islamic law.

Following are some Editions of his works:

- al-Rawḍ al-unuf fī šarḥ al-sīra al-Nabawiyya li-Ibn Hišām. wa-maʿahu al-Sīra al-Nabawiyya (7 volumes), 1967.
- al-Taʿrīf wa-al-iʿlām li-mā ubhima min al-Qurʾān min al-asmāʾ wa-al-aʿlām, Bayrut, 1987.
- Translation in German: Die Kommentare des Suhailī und des Abū Ḍarr zu den Uḥud-Gedichten in der Sīra des Ibn Hišam, Schaade, Arthur 1908.

Biographical Summary

Al-Suhayli was born in 1114 AD. He came to Marrakesh around 1182 at the call of the Almohad sultan Abu Yusuf Ya'qub al-Mansur. He died here three years later in 1185, and His zaouia is in a cemetery just outside Bab er Robb; it hides a former gate in the wall called Bab el Charia. The cemetery Bab Ech Charia, walled today, is built at the place where the Almohad troops of Abd El Moumen defeated the Almoravids in 1147.

43. Ibn Tufail

ʾAbū Bakr Muḥammad bin ʿAbd al-Malik bin Muḥammad ibn Ṭufayl al-Qaysiyy al-ʾAndalusiyy

(Arabic: أبو بكر محمد بن عبد الملك بن محمد بن طفيل القيسي الأندلسي),

(c. 1105 – 1185),

was a polymath from Andalusia; he was a writer, philosopher, theologian, physician, astronomer, and a vizier.

Contributions in Social Sciences

As a philosopher and novelist, ibn Ṭufayl is most famous for writing the first philosophical novel: Hayy ibn Yaqdhan.

As a physician, he was an early supporter of dissection and autopsy, which was expressed in his novel.

Ibn Tufail was the author of Ḥayy bin Yaqẓān (Arabic: حي بن يقظان), lit. 'Alive, son of Awake'), also known as Philosophus Autodidactus in Latin. It is a philosophical romance and allegorical novel inspired by Avicennism and Sufism. It tells the story of an autodidactic feral child, raised by a gazelle and living alone on a desert island, who, without contact with other human beings, discovers ultimate truth through a systematic process of reasoned inquiry.

Hayy ultimately comes into contact with civilization and religion when he meets a castaway named Absal (Asāl in some translations). He determines that certain trappings of religion, namely imagery and dependence on material goods, are necessary for the multitude in order that they might have decent lives. However, imagery and material goods are distractions from the truth and ought to be abandoned by those whose reason recognizes that they are.

The names of the characters in the novel, Ḥayy ibn Yaqẓān, Salamān, and Absāl were borrowed from Ibn Sina's tales. The title of the novel is also the same as Ibn Sina's novel. Ibn Tufail did this on purpose to use the characters and the title as a small reference to Ibn Sina, as he wanted to touch upon his philosophy.

Ibn Tufail's Philosophus Autodidactus was written as a response to al-Ghazali's book: The Incoherence of the Philosophers. In the 13th century, Ibn al-Nafis later wrote the Al-Risalah al-Kamiliyyah fil Siera al-Nabawiyyah (known as Theologus Autodidactus in the West), as a response to Ibn Tufail's Philosophus Autodidactus.

Hayy ibn Yaqdhan had a significant influence on both Arabic literature and European literature, and it went on to become an influential best-seller throughout Western Europe in the 17th and 18th centuries. The work also influenced both classical Islamic philosophy and modern Western philosophy. It became "one of the most important books that heralded the Scientific Revolution" and European Enlightenment.

The thoughts expressed in Hayy ibn Yaqdhan can be found in different variations and to different degrees in the books of Thomas Hobbes, John Locke, Isaac Newton, and Immanuel Kant.

A Latin translation of the work, entitled Philosophus Autodidactus, first appeared in 1671, prepared by Edward Pococke the Younger. The first English translation (by Simon Ockley) was published in 1708. These translations later may have inspired Daniel

Defoe to write Robinson Crusoe, which also featured a desert island narrative. The novel also inspired the concept of "tabula rasa" developed in An Essay Concerning Human Understanding (1690) by John Locke, who was a student of Pococke. His Essay went on to become one of the principal sources of empiricism in modern Western philosophy, and influenced many enlightenment philosophers, such as David Hume and George Berkeley.

Hayy ibn Yaqdhan's ideas on materialism in the novel also have some similarities to Karl Marx's historical materialism. It also foreshadowed Molyneux's Problem, proposed by William Molyneux to Locke, who included it in the second book of An Essay Concerning Human Understanding. Other European writers influenced by Philosophus Autodidactus included Gottfried Leibniz, Melchisédech Thévenot, John Wallis, Christiaan Huygens, George Keith, Robert Barclay, the Quakers, Samuel Hartlib, and Voltaire.

Another treatise by Ibn Tufayl is: Rağaz ṭawīl fī aṭ-Ṭibb (Arabic: رجز طويل في الطب), lit. 'Long Poem in Rajaz Metre on Medical Science'. It introduces the revolutionary ideas of Dissection and Autopsy.

Following are some sources of ibn Ṭufayl's works:

- Rağaz ṭawīl fī aṭ-Ṭibb (Arabic: رجز طويل في الطب, lit. 'Long Poem in Rajaz Metre on Medical Science': Is a long poem describing how to diagnose illnesses, and find their cures. The poem is written in the Arabic Rajaz metre. It was only found recently in Rabat, the capital of Morocco.
- Arabic text of Hayy bin Yaqzan from Wikisource.

- Full pdf of French translation of Hayy bin Yaqzan from Google Books.
- English translations of Hayy bin Yaqzan (in chronological order)
    - The improvement of human reason, exhibited in the life of Hai ebn Yokdhan, written in Arabic above 500 years ago, by Abu Jaafar ebn Tophail, newly translated from the original Arabic, by Simon Ockley. With an appendix, in which the possibility of man's attaining of the true knowledge of God, and things necessary to salvation, without instruction, is briefly considered. London: Printed and sold by E. Powell, 1708.
    - Abu Bakr Ibn Tufail, The history of Hayy Ibn Yaqzan, translated from the Arabic by Simon Ockley, revised, with an introduction by A.S. Fulton. London: Chapman and Hall, 1929. available online (omits the introductory section).
    - Ibn Tufayl's Hayy ibn Yaqzān: a philosophical tale, translated with introduction and notes by Lenn Evan Goodman. New York: Twayne, 1972.
    - The journey of the soul: the story of Hai bin Yaqzan, as told by Abu Bakr Muhammad bin Tufail, a new translation by Riad Kocache. London: Octagon, 1982.
    - Two Andalusian philosophers, translated from the Arabic with an introduction and notes by Jim Colville. London: Kegan Paul, 1999.

- Medieval Islamic Philosophical Writings, ed. Muhammad Ali Khalidi. Cambridge University Press, 2005. (omits the introductory section; omits the conclusion beginning with the protagonist's acquaintance with Absal; includes §§1-98 of 121 as numbered in the Ockley-Fulton version).
- Ben-Zaken, Avner, "Taming the Mystic", in Reading Hayy Ibn-Yaqzan: A Cross-Cultural History of Autodidacticism (Johns Hopkins University Press, 2011). ISBN 978-0801897399.

Biographical Summary

Born in Guadix, near Granada, he was educated by Ibn Bajjah (Avempace). His family were from the Arab Qays tribe. He was a secretary for several leaders, including the rulers of Ceuta and Tangier, in 1154. He also served as a secretary for the ruler of Granada, and later as vizier and physician for Abu Yaqub Yusuf, the Almohad caliph, to whom he recommended Ibn Rushd (Averroës) as his own future successor in 1169.

Ibn Rushd later reports this event and describes how Ibn Tufayl then inspired him to write his famous commentaries:

> Abu Bakr ibn Tufayl summoned me one day and told me that he had heard the Commander of the Faithful complaining about the disjointedness of Aristotle's mode of expression - or that of the translators - and the resultant obscurity of his intentions. He said that if someone took on these books who could summarize them and clarify their aims after first thoroughly understanding them

himself, people would have an easier time comprehending them. "If you have the energy," Ibn Tufayl told me, "you do it. I'm confident you can because I know what a good mind and devoted character you have, and how dedicated you are to the art. You understand that only my great age, the cares of my office - and my commitment to another task that I think even more vital - keep me from doing it myself."

Ibn Rushd became Ibn Tufayl's successor after he retired in 1182; Ibn Tufayl died several years later in Morocco in 1185.

The astronomer Nur Ed-Din Al-Bitruji was also a disciple of Ibn Tufayl. Al-Bitruji was influenced by him to follow astronomy.

Ibn Tufayl's work in astronomy was historically significant as he played a major role in overturning the Ptolemaic ideas on astronomy. This event in history is called the "Andalusian Revolt", where he influenced many, including Al-Bitruji, to desert the Ptolemaic ideas.

Many Islamic philosophers, writers, physicians, and astronomers have been influenced by Ibn Tufail and his work. These people include Nur al-Din al-Bitruji, Abu 'Abdallah Muhammad b. al-Abbar, Abd al-Wahid al-Marrakushi, Ahmed Mohammed al-Maqqari, and Ibn al-Khatib

44. Gerard of Cremona

Gerard of Cremona

(Latin: Gerardus Cremonensis),

(c. 1114 – 1187),

Was an Italian translator of scientific works from Arabic to Latin.

Contributions in Social Sciences

Gerard of Cremona's Latin translation of the Arabic version of Ptolemy's Almagest made c. 1175 was widely known in Western Europe before the Renaissance.

Gerard edited for Latin readers the Tables of Toledo, the most accurate compilation of astronomical data ever seen in Europe at the time. The Tables were partly the work of Al-Zarqali, known to the West as Arzachel, a mathematician and astronomer who flourished in Cordoba in the eleventh century.

Al-Farabi, the "second teacher" after Aristotle, wrote hundreds of treatises. His book on the sciences, Kitab lhsa al Ulum, discussed classification and fundamental principles of science in a unique and useful manner. Gerard rendered it as De scientiis (On the Sciences).

Gerard translated Euclid's Geometry and Alfraganus's Elements of Astronomy.

Gerard also composed treatises on algebra, arithmetic and astrology. In the astrology text, longitudes are reckoned from Toledo.

In total, Gerard of Cremona translated 87 books from the Arabic language, including such original Arabic works as al-Khwarizmi's On

Algebra and Almucabala, Jabir ibn Aflah's Elementa astronomica, and works by al-Razi (Rhazes).

The Latin translation of the Calendar of Córdoba, entitled Liber Anoe, has also been attributed to Gerard. However, some of the works credited to Gerard of Cremona are the works of a later Gerard Cremonensis, working in the thirteenth century, (who was also known as Gerard de Sabloneta (Sabbioneta)). The Europeans have confused the two translators with one another.

Other treatises translated by Gerard de Sabloneta include the Theoria or Theorica planetarum, and Avicenna's Canon of Medicine. This translation formed the basis of numerous subsequent Latin editions of Avicenna's works. Almansor of al-Razi was also translated.

Biographical Summary

Gerard of Cremona was a translator among the Toledo School of Translators who invigorated Europe by transmitting the Arabs' knowledge in astronomy, medicine and other sciences, by translating the knowledge into Latin.

Gerard was born in Cremona in northern Italy. Dissatisfied with the philosophies of his Italian teachers, Gerard went to Toledo. There he learned Arabic. Subsequently he went to Castile, no later than 1144.

Toledo had been a provincial capital in the Caliphate of Cordoba and remained a seat of learning and a multicultural capital, as its rulers promoted a large Christian, Jewish and Muslim community. The city was full of libraries and manuscripts, and was one of the few places in Europe where a Christian could be exposed to Arabic language and culture. In Toledo Gerard devoted the remainder of his life to making Latin translations from the Arabic scientific literature.

45. Usama ibn Munqidh

Majd ad-Dīn Usāma ibn Murshid ibn ʿAlī ibn Munqidh al-Kināni al-Kalbī

(Arabic: مجد الدّين أسامة ابن مُرشد ابن على ابن مُنقذ الكنانى الكلبى),

(July 4, 1095 – November 17, 1188),

was a poet, author, faris (knight), and diplomat.

Contributions in Social Sciences

During and immediately after his life, he was most famous as a poet and adib (a "man of letters"). He wrote many poetry anthologies, such as the Kitab al-'Asa ("Book of the Staff"), Lubab al-Adab ("Kernels of Refinement"), and al-Manazil wa'l-Diyar ("Dwellings and Abodes").

He wrote his own original poetry.

In modern times, he is remembered more for his Kitab al-I'tibar ("Book of Learning by Example" or "Book of Contemplation"), which contains lengthy descriptions of the crusaders, whom he interacted with on many occasions, and some of whom he considered friends.

Around 1171 in Hisn Kayfa, Usama wrote the Kitab al-'Asa ("Book of the Staff"), a poetry anthology about famous walking sticks and other staffs.

He also wrote al-Manazil wa'l-Diyar ("Dwellings and Abodes").

In Damascus in the early 1180s he wrote another anthology, the Lubab al-Adab ("Kernels of Refinement"), instructions on living a properly cultured life.

He is famous for the Kitab al-I'tibar (translated various ways, most recently as the Book of Contemplation), which was written as a gift to Saladin around 1183.

Philip Hitti translated the title as a "memoir", which it is not. It is not unusual of European writers to be sloppy in their works about Muslim scientists. It does include many autobiographical details because it was meant to be "a book of examples ('ibar) from which to draw lessons", though these details are only incidental to the main point."

In 1880, Hartwig Derenbourg discovered the Kitab al-I'tibar, which survived in only one manuscript, in the possession of the Escorial Monastery near Madrid. Derenbourg produced an Arabic edition (1886), a biography of Usama (1889), and a French translation (1895).

In 1930, Hitti produced an Arabic edition, and an English translation. Qasim as-Samarrai produced another Arabic edition in 1987.

Usama wrote in "Middle Arabic", a less formal style of classical Arabic. **Following are some Editions and translations of the works of** Usama ibn Munqidh:

- Ousama ibn Mounkidh, un emir Syrien au premier siècle des croisades (1095–1188), ed. Hartwig Derenbourg. Paris, 1889.
- *ibn Munqidh, Usama (1895).* Souvenirs historiques et récits de chasse *(in French).* Hartwig Derenbourg *(translator). Paris: E. Leroux.*

- *ibn Munqidh, Usama (1905).* Memoiren eines syrischen Emirs aus der Zeit der Kreuzzüge *(in German). Georg Schumann (translator). Innsbruck: Wagner'schen Universitäts - Buchhandlung.*
- *ibn Munqidh, Usama (1929).* An Arab-Syrian Gentleman And Warrior in The Period of The Crusades: Memoirs of Usama Ibn-Munqidh (Kitab al i'tibar). Philip K. Hitti *(translator). New York:* Columbia University Press.
- Memoirs Entitled Kitāb al-I'tibār, ed. Philip K. Hitti (Arabic text). Princeton: Princeton University Press, 1930.
- Lubab al-Adab, ed. A. M. Shakir. Cairo: Maktabat Luwis Sarkis, 1935.
- Diwan Usama ibn Munqidh, ed. A. Badawi and H. Abd al-Majid. Cairo: Wizarat al-Ma'arif al-Umumiyya, 1953.
- Kitab al-Manazil wa'l-Diyar, ed. M. Hijazi. Cairo: Al-Majlis al-A'la li-l-Shu'un al-Islamiyya, 1968.
- Kitab al-'Asa, ed. Hassan Abbas. Alexandria: Al-Hay'at al-Misriyya al-'Amma li-l-Kitab, 1978.
- Al-Badi' fi-l-Badi', ed. A. Muhanna. Beirut: Dar al-Kutub al-'Ilmiyya, 1987.
- Kitab al i'tibar, ed. Qasim as-Samarra'i. Riyadh, 1987.
- "Usama ibn Munqidh's Book of the Staff (Kitab al'Asa): autobiographical and historical excerpts," trans. Paul M. Cobb. Al-Masaq: Islam and the Medieval Mediterranean 17 (2005).

- "Usama ibn Munqidh's Kernels of Refinement (Lubab al-Adab): autobiographical and historical excerpts," trans. Paul M. Cobb. Al-Masaq: Islam and the Medieval Mediterranean 18 (2006)
- The Book of Contemplation: Islam and the Crusades, trans. Paul M. Cobb. Penguin Classics, 2008.

Following are some Secondary works of Usama ibn Munqidh:

- Ibn Khallikan's Biographical Dictionary, trans. William MacGuckin, Baron de Slane, vol. 1. Paris, 1842.
- Hassan Abbas, Usama ibn Munqidh: Hayatuhu wa-Atharuhu. Cairo: al-Hay'a al-Misriya al-'Ama li'l-Kitab, 1981.
- Adam M. Bishop, "Usama ibn Munqidh and crusader law in the twelfth century." Crusades 12 (2013), pp. 53-65.
- Niall Christie, "Just a bunch of dirty stories? Women in the memoirs of Usamah ibn Munqidh." Eastward Bound: Travel and Travellers, 1050–1550, ed. Rosamund Allen. Manchester: Manchester University Press, 2004, pp. 71–87.
- Paul M. Cobb, Usama ibn Munqidh: Warrior-Poet in the Age of Crusades Oxford: Oneworld, 2005.
- Paul M. Cobb, "Infidel dogs: hunting crusaders with Usamah ibn Munqidh." Crusades 6 (2007).
- Lawrence I. Conrad, "Usama ibn Munqidh and other witnesses to Frankish and Islamic medicine in the era of the crusades." Medicine in Jerusalem throughout the Ages, ed. Zohar Amar et al. Tel Aviv: C. G. Foundation, 1999.

- Carole Hillenbrand, The Crusades: Islamic Perspectives. Routledge, 2000.
- R. S. Humphreys, Munkidh, Banu. Encyclopaedia of Islam, 2nd. ed., vol. VII (Leiden: Brill, 1960–2002).
- Robert Irwin, "Usama ibn Munqidh: an Arab-Syrian gentleman at the time of the Crusades reconsidered." The Crusades and their sources: essays presented to Bernard Hamilton ed. John France, William G. Zajac (Aldershot: Ashgate, 1998) pp. 71–87.
- Adnan Husain, "Wondrous Crusade Encounters: Usamah ibn Munqidh's Book of Learning by Example," in Jason Glenn (ed), The Middle Ages in Texts and Texture: Reflections on Medieval Sources (Toronto, University of Toronto, 2012),
- D. W. Morray, "The genius of Usamah ibn Munqidh: aspects of Kitab al-I'tibar by Usamah ibn Munqidh." Working Paper. University of Durham, Centre for Middle Eastern and Islamic Studies, Durham, 1987.
- I. Schen, "Usama ibn Munqidh's Memoirs: some further light on Muslim Middle Arabic." Journal of Semitic Studies 17 (1972), and Journal of Semitic Studies 18 (1973).
- Bogdan C. Smarandache, "Re-examining Usama Ibn Munqidh's knowledge of "Frankish": A case study of medieval bilingualism during the crusades." The Medieval Globe 3 (2017), pp. 47-85.

- G. R. Smith, "A new translation of certain passages of the hunting section of Usama ibn Munqidh's I'tibar." Journal of Semitic Studies 26 (1981).
- Stefan Wild, "Open questions, new light: Usama ibn Munqidh's account of his battles against Muslims and Franks." The Frankish Wars and their Influence on Palestine, edd. Khalil Athamina and Roger Heacock (Birzeit, 1994), pp. 9–29.
- The Chronicle of Ibn al-Athir for the Crusading Period from al-Kamil i'l-Ta'rikh, Part 2: The Years 541–589/1146–1193: The Age of Nur al-Din and Saladin, trans. D.S. Richards. Crusade Texts in Translation 15. Aldershot: Ashgate, 2007.

Muslims know Usama ibn Munqidh for his poetry and his poetry anthologies. Ibn Khallikan, author of a fourteenth-century biographical dictionary, calls him "one of the most powerful, learned, and intrepid members of the [Munqidh] family" and speaks at great length about his poetry.

He was also known for his military and hunting exploits. Ibn al-Athir described him as "the ultimate of bravery", regarding his presence at the Battle of Harim.

For modern readers, he is known rather limitedly for his Kitab al-I'tibar and his descriptions of life in Syria during the early crusades. That is his contribution as a historian. The work is written with an anthological structure, with humorous or moralistic tales that are not meant to proceed chronologically. The European writers somewhat failed to understand his work, and some even thought that he was a rambler. But Usama ibn Munqidh was not writing a history, which the

European writers assumed that he was. For example, "adab" in Arabic does not necessarily have to be factual, European writers are quick to point out that Usama's historical material cannot always be trusted. As another example, Usama's anecdotes about the crusades are sometimes obvious jokes, exaggerating their "otherness" to entertain his Muslim audience; but European writer Carole Hillenbrand misunderstood the book and blurted that it would be "dangerously misleading to take the evidence of his book at its face value."

Biographical Summary

His life coincided with the rise of several dynasties, the arrival of the First Crusade, and the establishment of the crusader states.

He was the nephew and potential successor of the emir of Shaizar, but was exiled in 1131 and spent the rest of his life serving other leaders. He was a courtier to the Burids, Zengids, and later Ayyubids in Damascus, serving Nur ad-Din Zengi, and Saladin, over a period of almost fifty years. He also served the Fatimid court in Cairo, as well as the Artuqids in Hisn Kayfa.

He travelled extensively in Arab lands, visiting Egypt, Syria, Palestine, and along the Tigris River. He also went on pilgrimage to Mecca.

He often meddled in the politics of the courts in which he served, and he was exiled from both Damascus and Cairo.

Most of his family was killed in an earthquake at Shaizar in 1157. He died in Damascus in 1188, at the age of 93.

Usama was the son of Murshid, and the nephew of Nasr, emir of Shaizar. Shaizar was seen as a strategically important site and the gateway to enter and control inner Syria. The Arabs initially conquered Shaizar during the Muslim conquest of the Levant in 637. Due to its importance, it exchanged hands numerous times between the Arabs and Byzantines.

In 1025 the Banu Munqidh tribe were given an allocation of land beside Shaizar by the ruler of Hama, Salih ibn Mirdas. Over time they expanded their lands, building fortifications and castles until Usama's grandfather Izz al-Dawla al-Murhaf Nasr retook it in 1080.

When Nasr died in 1098, Usama's father, Majd ad-Din Abi Salamah Murshid (1068-1137) became the emir of Shaizar and the surrounding cities. However, Murshid was more interested in studying religion and hunting than in matters of statecraft. So Usama's father, Murshid, gave up his position to Usama's uncle, Izz ad-Din Abi al-Asaker Sultan.

During the rule of Usama's uncle, Shaizar was attacked numerous times by the Banu Kilab of Aleppo, the sect of the Hashshashin, the Byzantines, and the crusaders. It was struck with siege engines for 10 days in 1137 by the Byzantines and the crusaders attempted on many occasions to storm it. However, due to its fortifications, it never fell.

As a child, Usama was the second of four boys and raised by his nurse, Lu'lu'a, who had also raised his father and would later raise Usama's own children. He was encouraged by his father to memorize the Quran, and was also tutored by scholars such as Ibn Munira of Kafartab and Abu Abdullah al-Tulaytuli of Toledo. He spent much of his youth hunting with his family, partly as recreation and certainly as warrior (faris), training for battle as part of furusiyya. He also gathered much direct fighting experience, against the neighboring crusader County of Tripoli and Principality of Antioch, hostile Muslim neighbors in Hama, Homs, and elsewhere, and against the Hashshashin who had established a base near Shaizar.

Sultan did not initially have any male heirs and it is possible that Usama expected to succeed him. He certainly singled him out among his brothers by teaching him, tutoring him in the ways of war and hunting. He even favored him for personal missions and as a representative.

However, after Sultan had his own son, he no longer appreciated the presence of Usama and Murshid's other sons. According to Usama, Sultan became jealous after a particularly successful lion-hunt in 1131, when Usama entered the town with a large lion head in his arms as a hunting trophy. When his grandmother saw this, she warned him about the effect this could have on his uncle.

Despite this, he still spoke well of his uncle on a few occasions in his autobiography and highlighted his noble actions. Usama ultimately left Shaizar temporarily in 1129, and after his father's death in 1137 his exile became permanent.

Usama's uncle died in 1154 and his son, Taj al-Dawla Nasr ad-Din Muhammad, inherited the Shaizar state. However, Usama was the last heir of the line left alive when in 1157 an earthquake struck the area, killing most of his family.

Usama went to Homs, where he was taken captive in a battle against Zengi, the atabeg of Mosul and Aleppo, who had just captured nearby Hama. After his capture Usama ibn Munqidh entered Zengi's service, and travelled throughout northern Syria, Iraq, and Armenia fighting against Zengi's enemies; including the Abbasid caliph outside Baghdad in 1132. In 1135, he returned to Hama, where one of Zengi's generals, al-Yaghisiyani, was appointed governor.

Usama ibn Munqidh returned to Shaizar when his father died in May 1137, and again in April 1138 when Byzantine emperor John II Comnenus besieged the city. The emperor's siege of Shaizar was unsuccessful, but Shaizar was heavily damaged. After the siege, Usama left Zengi's service and went to Damascus, which was ruled by Mu'in ad-Din Unur, the atabeg of the Burid dynasty. Zengi was determined to conquer Damascus, so Usama and Unur turned to the crusader Kingdom of Jerusalem for help. Usama was sent on a preliminary visit to Jerusalem in 1138. In 1139 Zengi captured Baalbek in Damascene territory. In 1140, Unur sent Usama back to Jerusalem to conclude a treaty with the crusaders, and both he and Unur visited their new allies, numerous times between 1140 and 1143.

During these diplomatic missions Usama developed a friendship with members of the Knights Templar whom he considered more civilized than other crusader orders.

Afterwards, Usama was suspected of being involved in a plot against Unur, and he fled Damascus for Fatimid Cairo in November, 1144.

In Cairo he became a wealthy courtier, but he was involved in plots and conspiracies there as well. The young az-Zafir became caliph in 1149, and Ibn as-Sallar became vizier, with Usama as one of his advisors. As-Sallar sent Usama to negotiate an alliance against the crusaders with Zengi's son Nur ad-Din, but the negotiations failed.

Usama took part in battles with the crusaders outside of Ascalon on his way back to Egypt; and after he left, his brother 'Ali was killed at Gaza.

Back in Egypt, as-Sallar was assassinated in 1153 by his son Abbas, Abbas's son Nasr, and caliph az-Zafir, who, according to Usama, was Nasr's lover. Thirteenth-century historian Ibn al-Athir says that Usama was the instigator of this plot.

Az-Zafir's relatives called upon a supporter, Tala'i ibn Ruzzik, who chased Abbas out of Cairo, and Usama followed him. He lost his possessions in Cairo, and on the way to Damascus his retinue was attacked by the crusaders and Bedouin nomads; however, in June 1154 he safely reached Damascus, which had recently been captured by Nur ad-Din.

Ibn Ruzzik tried to persuade Usama ibn Munqidh to come back, as the rest of his family was still in Cairo, but Usama was able to bring his family to Damascus, through crusader territory, in 1156.

The crusaders promised to transport his family safely, but they were attacked and pillaged.

In 1157, Shaizar was destroyed by an earthquake, killing almost all of Usama's relatives. They were there for the circumcision of the son of his cousin Muhammad, who had recently succeeded Sultan as emir. The only survivor was Muhammad's wife. Usama had remained in Damascus, and after the destruction of his homeland he remained there in semi-retirement. He went on pilgrimage to Mecca in 1160, then went on campaign against the crusaders with Nur ad-Din in 1162, and was at the Battle of Harim in 1164. That year, Usama left Nur ad-Din's service and went north to the court of Kara Arslan, the Artuqid emir of Hisn Kayfa.

Usama's life in Hisn Kayfa is very obscure, but he travelled throughout the region, and probably wrote many of his works there. In 1174, Usama was invited to Damascus to serve Saladin, who had succeeded Nur ad-Din earlier that year and was a friend of Usama's son Murhaf. Usama lived in semi-retirement, as he did in Hisn Kayfa, and often met with Saladin to discuss literature and warfare.

Usama may have also taught poetry and hadith in Damascus, and held poetry salons for Saladin and his chief men, including al-Qadi al-Fadil and Imad ad-Din al-Isfahani.

He died on November 17, 1188. He was buried in Damascus on Mount Qasiyun, although the tomb is now lost.

Usama had three brothers, Muhammad, 'Ali, and Munqidh. His cousin, also named Muhammad, succeeded Usama's uncle Sultan as emir of Shaizar. He had a son, Murhaf, in 1126, and another son, Abu Bakr, who died as a child. He had a daughter, Umm Farwa, in Hisn Kayfa in 1166. He mentions other children, but their names are unknown.

The picture he painted of his father was of a pious religious man who was not interested in the affairs of this world. He would spend most of his time reading the Quran, fasting and hunting during the day and at night would copy the Quran. He also recounted a few battles his father joined against the crusaders in his Kitab al Itibar.

46. Ibn Mada

Abu al-Abbas Ahmad bin Abd al-Rahman bin Muhammad bin Sa'id bin Harith bin Asim al-Lakhmi al-Qurtubi,

also known as Ibn Maḍāʾ (Arabic: ابن مضاء),

(1116–1196),

was a polymath, mathematician and grammarian, from Córdoba in Islamic Spain.

Contributions in Social Sciences

Ibn Mada was notable for having challenged the traditional formation of Arabic grammar and of the common understanding of linguistic governance among Arab grammarians, performing an overhaul that was first suggested by Al-Jahiz 200 years prior. He addressed the subject of dependency in the grammatical sense in which it is understood today.

In addition to Arabic grammar and religion, he was well-versed in geometry and medicine.

Ibn Mada rose to fame as one of the first to launch attack on Arabic grammar theory and called for its reformation. Although he was concerned with attacking all major schools of Arabic grammar, he was focused on the grammar of the linguists of Basra, as it was the most popular school around him. His attack on eastern Arabic grammar was reasoned and eloquent, defending his view that grammar as it was understood in that region was complicated, casuistic, obscure and artificial. Ibn Mada, instead called for building simple and clear grammar based on true facts of the language. His ideas were considered revolutionary during his life, as among his ideas was the abolition of governance and linguistic analogy. Ibn Mada felt that scholarly work on the Arabic language was intentionally convoluted and inaccessible to both non-native speakers and laymen Arabs, and that an overall simplification of language and grammar would enhance overall comprehension of Arabic language. Ibn Mada held great respect for the language as the native speakers understood it, and while he emphasized a simplification of grammar, he did not advocate a complete overhaul of the entire language.

There was renewed interest in his work during the 1950s.

His Zahirite views in Muslim jurisprudence influenced his views in linguistics. He explicitly denied the ability of human beings to willfully choose what they say and how they say it, since speech – like all other things – is predetermined by God.

Because Arabic grammarians during Ibn Mada's time often linked the spoken language to grammatical causes, they earned both his linguistic and theological ire. He was thus influential and instrumental during the Almohad reforms as chief judge of the Almohad Caliphate.

In his view, the Zahirite denial of legal causality in regard to Islamic law carried over into a denial of linguistic causality in regard to Arabic grammar.

Ibn Mada's reaction toward Arabic grammar and grammarians wasn't without provocation. Both earlier Zahirite jurists such as Ibn Hazm and al-Ballūṭī and some Shafi'ites sparred with Hanafite jurists who sought to justify practices such as Istihsan, anathema to the more orthodox schools, on the basis of grammatical and linguistic arguments. Thus, suspicion and antagonism toward grammarians in the east, where the Hanafite rite predominated, had already been started before Ibn Mada began his whole-scale effort.

Ibn Mada's mastery of the Arabic language and its subfields was so great that, at the time, he was said to have been isolated from the general body of scholarship in terms of sheer knowledge. His refutation was written toward the end of his life and demonstrated his clarity of thought and independent judgment, causing his student Ibn Dihya al-Kalby to brand him as the leader of all grammarians. His critical views of Arabic grammar as it was taught in the east found an audience with other linguistic and religious scholars of the western half, Abu Hayyan Al Gharnati being one example.

Gharnati also criticized so-called "eastern grammarians" and his treatise on the non-existence of grammatic causality cited Ibn Mada as his inspiration. Although Ibn Dihya, Abu Hayyan and Ibn Mada shared their Zahirite and Andalusian backgrounds, not all of Ibn Mada's intellectual descendants shared these traits.

Thus, while Ibn Mada opened the discussion regarding the competence of grammarians, suspicion surrounding them and the religious implications of their work continued even after his death.

In the mid-20th century, the rediscovery of Ibn Mada's Refutation by Egyptian linguist Shawqi Daif caused minor shockwaves. Convinced that Ibn Mada's abolition of linguistic analogy and governance were the solutions to the failure of Arabic language education, Daif used this foundation for his later advocacy of modernizing language arts education in the Middle East.

Following are some works of Ibn Mada:

- al-Radd 'alá al-Nuḥāh (الرد على النحاة)
- al-Mashriq fī Iṣlāḥ al-Manṭiq (المشرق في إصلاح المنطق)
- Tanzīh al-Qur'ān 'ammā lā Yalīqu min al-Bayān (تنزيه القرآن عما لا يليق من البيان)

Biographical Summary

Ibn Mada's exact date of birth has been listed as both 1116 and 1119 AD. His family was famous within their local community. Ibn Mada was not known to have traveled outside of Cordoba prior to his academic study. He grew up in a family of noble origin, and as a youth he seemed to concern himself with pursuing his education.

He moved from Cordoba to Seville where he studied Arabic grammar and syntax from the works of Sibawayh. Later, he left the Iberian Peninsula for Ceuta in North Africa in order to study historiography and prophetic tradition with Muslim academic Qadi Ayyad. Ibn Mada was most affected by his linguistic study, excelling so far as to develop his own independent opinions in regard to disputes among grammarians.

Ibn Mada initially served as a judge in Fes, in present-day Morocco, and later at Béjaïa in present-day Algeria. It was during his initial judgeship that he was a teacher of fellow Andalusian theologian and litterateur, Ibn Dihya al-Kalby.

Later on, Almohad Caliph Abu Yaqub Yusuf chose him to serve as the chief judge for the caliphate. He served in Fes, Marrakesh and

Seville, outliving Abu Yaqub to serve under the caliph's son Abu Yusuf Yaqub al-Mansur and remaining as the empire's chief judge for the remainder of his life. During the Almohad reforms, he assisted the Almohad authorities in banning any and all religious books written by non-Zahirites, during the reign of Abu Yaqub Yusuf, and oversaw the outright burning of such books under Yusuf's son Abu Yusuf Yaqub al-Mansur.

He died in Seville the Islamic calendar month of Jumada al-awwal during the Hijri year of 592 corresponding to 1196 in the Gregorian calendar, just as he was approaching eighty years of age.

47. Ibn Rushd (Averroes)

Abū l-Walīd Muḥammad Ibn ʾAḥmad Ibn Rushd

(Arabic: أبو الوليد محمد ابن احمد ابن رشد), often Latinized as Averroes,

(14 April 1126 – 11 December 1198),

was an Andalusian polymath who wrote about many subjects, including philosophy, theology, medicine, astronomy, physics, psychlogy, mathematics, Islamic jurisprudence, law, and linguistics.

Ibn Rushd taught Rationalism to the Western world, who regards him as the Father of Rationalism.

Contributions in Social Sciences

In Islamic jurisprudence, Ibn Rushd wrote the Bidāyat al-Mujtahid on the differences between Islamic schools of law and the principles that caused their differences.

In medicine, Ibn Rushd proposed a new theory of stroke, described the signs and symptoms of Parkinson's disease for the first time, and might have been the first to identify the retina as the part of the eye responsible for sensing light. His medical book Al-Kulliyat fi al-Tibb, translated into Latin and known as the Colliget, became a textbook in Europe for centuries.

In the West, Ibn Rushd was known for his extensive commentaries on Aristotle, many of which were translated into Latin and Hebrew. The translations of his work reawakened western European interest in Aristotle and Greek thinkers, an area of study that had been widely abandoned after the fall of the Western Roman Empire. His thoughts generated controversies in Latin Christendom and triggered a philosophical movement called Averroism based on his writings. His unity of the intellect thesis, proposing that all humans share the same intellect, became one of the best-known and most controversial Averroist doctrines in the West. His works were condemned by the Catholic Church in 1270 and 1277. Although weakened by condemnations and sustained critique from Thomas Aquinas, Latin Averroism continued to attract followers up to the sixteenth century.

"Averroes" is the Medieval Latin form of "Ibn Rushd". Other forms of the name in European languages include "Ibin-Ros-din", "Filius Rosadis", "Ibn-Rusid", "Ben-Raxid", "Ibn-Ruschod", "Den-Resched", "Aben-Rassad", "Aben-Rasd", "Aben-Rust", "Avenrosdy", "Avenryz", "Adveroys", "Benroist", "Avenroyth" and "Averroysta".

According to French author Ernest Renan, Ibn Rushd wrote at least 67 original works, including 28 works on **philosophy**, 20 on **medicine**, 8 on **law**, 5 on **theology**, and 4 on **grammar**. He expressed his original contributions also in his commentaries on Aristotle's works, like Physics, Metaphysics, On the Soul, On the Heavens, and Posterior Analytics; and The Republic by Plato. Many of Ibn Rushd's works in Arabic did not survive, except their translations into Hebrew or Latin.

Ibn Rushd wrote stand-alone **philosophical treatises**, including On the Intellect, On the Syllogism, On Conjunction with the Active Intellect, On Time, On the Heavenly Sphere and On the Motion of the Sphere.

He wrote several polemics: Essay on al-Farabi's Approach to Logic, Metaphysical Questions Dealt with in the Book of Healing by Ibn Sina, and Rebuttal of Ibn Sina's Classification of Existing Entities.

Fakhry has named three works as Ibn Rushd's key writings in **theology.** Fasl al-Maqal ("The Decisive Treatise") is an 1178 treatise that argues for the compatibility of Islam and philosophy. Al-Kashf 'an Manahij al-Adillah ("Exposition of the Methods of Proof"), written in 1179, criticizes the theologies of the Asharites, and lays out Ibn Rushd's argument for proving the existence of God, as well as his thoughts on God's attributes and actions. The 1180 Tahafut al-Tahafut ("Incoherence of the Incoherence") is a rebuttal of al-Ghazali's (d. 1111) landmark criticism of philosophy, The Incoherence of the Philosophers. It combines his works, and uses them to respond to al-Ghazali.

Ibn Rushd, who served as the royal physician at the Almohad court, wrote a number of **medical treatises.** Al-Kulliyat fi al-Tibb ("The General Principles of Medicine"), was written around 1162, before his appointment at court. Its Latin translation, the Colliget, became a medical textbook in Europe for centuries.

His friend Ibn Zuhr wrote al-Juz'iyyat fi al-Tibb ("The Specificities of Medicine"). The two authors collaborated, intending that their works complement each other.

His other surviving titles include On Treacle, The Differences in Temperament, and Medicinal Herbs. He wrote a commentary on Avicenna's Urjuzah fi al-Tibb ("Poem on Medicine").

Ibn Rushd served multiple tenures as judge and produced multiple works in the fields of Islamic **jurisprudence** or legal theory. The only book that survives today is Bidāyat al-Mujtahid wa Nihāyat al-Muqtaṣid ("Primer of the Discretionary Scholar"). In this work he explains the differences of opinion (ikhtilaf) between the Sunni madhhabs (schools of Islamic jurisprudence) both in practice and in their underlying juristic principles, as well as the reason why they are inevitable. Despite his status as a Maliki judge, the book also discusses the opinion of other schools, including liberal and conservative ones. Other than this surviving text, bibliographical information shows he wrote a summary of Al-Ghazali's On Legal Theory of Muslim Jurisprudence (Al-Mustasfa) and tracts on sacrifices and land tax.

He rejected Avicenna's modality and Avicenna's argument to prove the existence of God as the Necessary Existent.

During Ibn Rushd's lifetime, philosophy came under attack from the Sunni Islam tradition, especially from theological schools like the traditionalist (Hanbalite) and the Ashari schools. In particular, the Ashari scholar al-Ghazali (1058–1111) wrote The Incoherence of the Philosophers (Tahafut al-falasifa), a scathing and influential critique of the philosophical tradition in the Islamic world. Among others, Al-Ghazali charged philosophers with non-belief in Islam and sought to disprove the teaching of the philosophers using logical arguments.

In Decisive Treatise, Ibn Rushd argues that philosophy cannot contradict revelations in Islam because they are just two different methods of reaching the truth, and "truth cannot contradict truth". When conclusions reached by philosophy appear to contradict the text of the revelation, then according to Ibn Rushd, revelation must be subjected to interpretation or allegorical understanding to remove the contradiction. This interpretation must be done by those "rooted in knowledge" - a phrase taken from the Quran, 3:7, which for Ibn Rushd refers to philosophers who during his lifetime had access to the "highest methods of knowledge". He also argues that the Quran calls for Muslims to study philosophy because the study and reflection of nature would increase a person's knowledge of "the Artisan" (God). He quotes Quranic passages calling on Muslims to reflect on nature and uses them to render a fatwa (legal opinion) that philosophy is allowed for Muslims and is probably an obligation, at least among those who have the talent for it.

Ibn Rushd also distinguishes between three modes of discourse: the rhetorical (based on persuasion) accessible to the common masses; the dialectical (based on debate) and often employed by theologians and the ulama (scholars); and the demonstrative (based on logical deduction). According to Ibn Rushd, the Quran uses the rhetorical method of inviting people to the truth, which allows it to reach the common masses with its persuasiveness, whereas philosophy uses the demonstrative methods that were only available to the learned but provided the best possible understanding and knowledge.

Ibn Rushd lays out his views on the existence and nature of **God** in the treatise The Exposition of the Methods of Proof. He examines and critiques the doctrines of four sects of Islam: the Asharites, the Mutazilites, the Sufis and those he calls the "literalists" (al-hashwiyah). Among other things, he examines their proofs of God's existence and critiques each one. Ibn Rushd argues that there are two arguments for God's existence that he deems logically sound and in accordance to the Quran; the arguments from "providence" and "from invention". The providence argument considers that the world and the universe seem finely tuned to support human life. Ibn Rushd cited the sun, the moon, the rivers, the seas and the location of humans on the earth. According to him, this suggests a creator who created them for the welfare of mankind. The argument from invention contends that worldly entities such as animals and plants appear to have been invented. Therefore, Ibn Rushd argues that a designer was behind the creation and that is God. Ibn Rushd's two arguments are teleological in nature and not cosmological like the contemporaneous Muslim kalam theologians.

Ibn Rushd upholds the doctrine of divine unity (tawhid) and argues that God has seven divine **attributes**: knowledge, life, power, will, hearing, vision and speech. He devotes the most attention to the attribute of knowledge and argues that divine knowledge differs from human knowledge because God knows the universe because God is its cause while humans only know the universe through its effects.

Ibn Rushd argues that the attribute of life can be inferred because it is the precondition of knowledge and also because God willed objects into being. Power can be inferred by God's ability to bring creations into existence. Ibn Rushd also argues that knowledge and power inevitably give rise to speech. Regarding vision and speech, he says that because God created the world, he necessarily knows every part of it in the same way an artist understands his or her work intimately. Because two elements of the world are the visual and the auditory, God must necessarily possess vision and speech.

The omnipotence paradox was first addressed by Ibn Rushd and only later by Thomas Aquinas.

In the centuries preceding Ibn Rushd, there had been a debate between Muslim thinkers questioning whether the world was created at a specific moment in time or whether it has always existed. Neo-Platonic **philosophers** such as Al-Farabi and Avicenna argued the world has always existed. This view was criticized by theologians and philosophers of the Ashari kalam tradition; in particular, al-Ghazali wrote an extensive refutation of the pre-eternity doctrine in his Incoherence of the Philosophers and accused the Neo-Platonic philosophers of unbelief (kufr).

Ibn Rushd responded to Al-Ghazali in his Incoherence of the Incoherence. First, he argued that the differences between the two positions were not vast enough to warrant the charge of unbelief. He also said the pre-eternity doctrine did not necessarily contradict the Quran and cited verses that mention pre-existing "throne" and "water" in passages related to creation. Ibn Rushd argued that a careful reading of the Quran implied only the "form" of the universe was created in time but that its existence has been eternal. Ibn Rushd further criticized the kalam theologians for using their own interpretations of scripture to answer questions that should have been left to philosophers.

Ibn Rushd states his **political philosophy** in his commentary of Plato's Republic. He combines his ideas with Plato's and with Islamic tradition; he considers the ideal state to be one based on the Islamic law. His interpretation of Plato's philosopher-king followed that of Al-Farabi, which equates the philosopher-king with the imam, caliph and lawgiver of the state. Ibn Rushd's description of the characteristics of a philosopher-king are similar to those given by Al-Farabi; they include love of knowledge, good memory, love of learning, love of truth, dislike for sensual pleasures, dislike for amassing wealth, magnanimity, courage, steadfastness, eloquence and the ability to "light quickly on the middle term". Ibn Rushd writes that if philosophers cannot rule - as was the case in the Almoravid and Almohad empires around his lifetime - philosophers must still try to influence the rulers towards implementing the ideal state.

According to Ibn Rushd, there are two methods of teaching virtue to citizens; persuasion and coercion. Persuasion is the more natural method consisting of rhetorical, dialectical and demonstrative methods; sometimes, however, coercion is necessary for those not amenable to persuasion, e.g. enemies of the state. Therefore, he justifies war as a last resort, which he also supports using Quranic arguments. Consequently, he argues that a ruler should have both wisdom and courage, which are needed for governance and defense of the state.

Like Plato, Ibn Rushd calls for women to share with men in the administration of the state, including participating as soldiers, philosophers and rulers. He regrets that contemporaneous Muslim societies limited the public role of women; he says this limitation is harmful to the state's well-being.

Ibn Rushd also accepted Plato's view of the deterioration of the ideal state. He cites examples from Islamic history when the Rashidun caliphate - which in Sunni tradition represented the ideal state led by "rightly guided caliphs" - became a dynastic state under Muawiyah, founder of the Umayyad dynasty. He also says the Almoravid and the Almohad empires started as ideal, shariah-based states but then deteriorated into timocracy, oligarchy, *democracy* and tyranny.

In his tenure as judge and jurist, Ibn Rushd for the most part ruled and gave fatwas according to the Maliki school of Islamic law which was dominant in Al-Andalus and the western Islamic world during his time. However, he frequently acted as "his own man", including sometimes rejecting the "consensus of the people of Medina" argument that is one of the traditional Maliki positions. In Bidāyat al-Mujtahid, one of his major contributions to the field of **Islamic law**, he not only describes the differences between various school of Islamic laws but also tries to theoretically explain the reasons for the difference and why they are inevitable. Even though all the schools of Islamic law are ultimately rooted in the Quran and hadith, there are "causes that necessitate differences" (al-asbab al-lati awjabat al-ikhtilaf). They include differences in interpreting scripture in a general or specific sense, in interpreting scriptural commands as obligatory or merely recommended, or prohibitions as discouragement or total prohibition, as well as ambiguities in the meaning of words or expressions. Ibn Rushd also writes that the application of qiyas (reasoning by analogy) could give rise to different legal opinion because jurists might disagree on the applicability of certain analogies and different analogies might contradict each other.

As did Ibn Bajja (Avempace) and Ibn Tufail, Ibn Rushd criticizes the Ptolemaic **astronomical** system using philosophical arguments and rejects the use of eccentrics and epicycles to explain the apparent motions of the Moon, the Sun and the planets. He argued that those objects move uniformly in a strictly circular motion around the Earth. He postulates that there are three type of planetary motions; those that can be seen with the naked eye, those that requires instruments to observe and those that can only be known by philosophical reasoning. Ibn Rushd argues that the occasional opaque colors of the Moon are caused by variations in its thickness; the thicker parts receive more light from the Sun and therefore emit more light than the thinner parts. This explanation was used up to the seventeenth century by the European Scholastics to account for Galileo's observations of spots on the Moon's surface, until the Scholastics such as Antoine Goudin in 1668 conceded that the observation was more likely caused by mountains on the Moon.

Ibn Rushd (Averroes) and Ibn Bajja (Avempace) observed sunspots, which they thought were transits of Venus and Mercury between the Sun and the Earth. In 1153 Ibn Rushd conducted astronomical observations in Marrakesh, where he observed the star Suhayl (Canopus) which was invisible in the latitude of his native Spain. He used this observation to support Aristotle's argument for the spherical Earth.

Ibn Rushd was aware that Arabic and Andalusian astronomers of his time focused on "mathematical" astronomy, which enabled accurate predictions through calculations but did not provide a detailed physical explanation of how the universe worked. According to him, "the astronomy of our time offers no truth, but only agrees with the calculations and not with what exists." He attempted to reform astronomy to be reconciled with physics. His works influenced astronomer Nur ad-Din al-Bitruji (d. 1204) who adopted most of his reform principles and did succeed in proposing an early astronomical system.

In **physics**, Ibn Rushd did not adopt the inductive method that was developed by Al-Biruni and is closer to today's physics. Rather, he was an "exegetical" scientist who produced new theses about nature through discussions of previous texts. Ibn Rushd's work introduced highly original theories of physics, especially his elaboration on motion as forma fluens/fluxus formae. These were taken up in the west as important to the overall development of physics.

Ibn Rushd also proposed a definition of force as "the rate at which work is done in changing the kinetic condition of a material body".

Ibn Rushd expounds his thoughts on **psychology** in his interest in explaining the human intellect using philosophical methods. In his short commentary, the first of the three works, Ibn Rushd follows Ibn Bajja's theory that something called the "material intellect" stores specific images that a person encounters. These images serve as basis for the "unification" by the universal "agent intellect", which, once it happens, allows a person to gain universal knowledge about that concept. In his middle commentary, Ibn Rushd moves towards the ideas of Al-Farabi and Ibn Sina (Avicenna), saying the agent intellect gives humans the power of universal understanding, which is the material intellect. Once the person has sufficient empirical encounters with a certain concept, the power activates and gives the person universal knowledge (see also logical induction). In his last commentary, called the Long Commentary, he proposes another theory, which becomes known as the theory of "the unity of the intellect". In it, Ibn Rushd argues that there is only one material intellect, which is the same for all humans and is unmixed with human body. To explain how different individuals can have different thoughts, he uses a concept he calls fikr, a process that happens in human brains and contains not universal knowledge but "active consideration of particular things" the person has encountered.

While his works in **medicine** indicate an in-depth theoretical knowledge in medicine, in one of his works he stated that he had not "practiced much apart from myself, my relatives or my friends."

However, he did serve as a royal physician.

Ibn Rushd's medical work is Al-Kulliyat fi al-Tibb. Ibn Rushd's original contributions include his observations on the retina. Ibn Rushd was the first to recognize that retina was the part of the eye responsible for sensing light. In his time, it was commonly thought that the lens was responsible for sensing light. This was a departure from Galen.

Western scholars often hold back such acknowledgements where the Muslim scientists were seen as expanding the frontiers of science. In this case, they try to create doubt whether this is what Ibn Rushd meant in his Kulliyat. However, such maneuvering in the present case is falsified by the fact that, in another treatise, Ibn Rushd observed that "the innermost of the coats of the eye [the retina] must necessarily receive the light from the humors of the eye [the lens], just like the humors receive the light from air."

Another of his departure from Galen and the medical theories of his time is his description of stroke as produced by the brain and caused by an obstruction of the arteries from the heart to the brain. This explanation reflects the modern understanding of the disease, which also attributes it to the obstruction between heart and the periphery.

Ibn Rushd, in his Kulliyat, was also the first to describe the signs and symptoms of Parkinson's disease.

The works of Ibn Rushd influenced the future civilizations.

Thirteenth century Jewish researchers, Maimonides, Samuel ibn Tibbon, Judah ibn Solomon Cohen, and Shem-Tov ibn Falaquera relied heavily on Ibn Rushd's work.

In 1232, Joseph Ben Abba Mari translated Ibn Rushd's commentaries, which was the first Jewish translation of a complete work. In 1260 Moses ibn Tibbon published the translation of almost all of Ibn Rushd's commentaries and some of his works on medicine. Ibn Rushd's influence peaked in the fourteenth century, including Kalonymus ben Kalonymus of Arles, Todros Todrosi of Arles, Elia del Medigo of Candia and Gersonides of Languedoc.

Ibn Rushd's main influence on the Christian West was through his extensive commentaries on Aristotle. After the fall of the Western Roman Empire, western Europe fell into a cultural decline. Ibn Rushd's researches were translated into Latin and entered western Europe in the thirteenth century. He is regarded as the "father of free thought" and "father of rationalism". Michael Scot (1175 – c. 1232) was the first Latin translator of Ibn Rushd who translated the long commentaries of Physics, Metaphysics, On the Soul and On the Heavens, starting in 1217 in Paris and Toledo. Other European authors such as Hermannus Alemannus, William de Luna and Armengaud of Montpellier translated Ibn Rushd's other works, sometimes with help from Jewish authors. Ibn Rushd's works propagated among Christian scholastic tradition. His writing attracted a strong circle of followers. Paris and Padua were major centers of Latin Averroism, and its prominent thirteenth-century leaders included Siger of Brabant and Boethius of Dacia.

Authorities of the Roman Catholic Church reacted against the spread of Ibn Rushd (Averroism). In 1270, the Bishop of Paris Étienne Tempier issued a condemnation against 15 doctrines, many of which were from Ibn Rushd. He said they were in conflict with the doctrines of the church. In 1277, at the request of Pope John XXI, Tempier issued another condemnation, this time targeting 219 theses drawn from many sources, mainly the teachings of Ibn Rushd.

Ibn Rushd also received a mixed reception from individual Catholic thinkers. Thomas Aquinas, a leading Catholic thinker of the thirteenth century, relied extensively on Ibn Rushd's works though he disagreed with him on many points. For example, he wrote a detailed attack on Ibn Rushed's theory that all humans share the same intellect. He also opposed Ibn Rushd on the eternity of the universe and divine providence.

The Catholic Church's condemnations of 1270 and 1277, and the detailed critique by Aquinas weakened the spread of Ibn Rushd's influence in Latin Christendom, though it maintained a following until the sixteenth century. Leading followers of Ibn Rushd in the following centuries included John of Jandun and Marsilius of Padua (fourteenth century), Gaetano da Thiene and Pietro Pomponazzi (fifteenth century), and Agostino Nifo and Marcantonio Zimara (sixteenth century).

Ibn Rushd did not have a major influence on Islamic philosophic thought because there were so many great scientists in the Muslim world, unlike Europe that was trying to emerge from its Dark Ages. Also, Spain was in the extreme west of the Islamic civilization, far from the centers of Islamic intellectual traditions where there were so many greater luminaries. They were engaging deeply with newer schools of thought, like that of Ibn Sina.

Ibn Rushd is depicted in Raphael's 1501 fresco The School of Athens that decorates the Apostolic Palace in the Vatican, which features seminal figures of philosophy. In the painting, Averroes (Ibn Rushd) wears a green robe and a turban, and peers out from behind Pythagoras, who is shown writing a book.

Biographical Summary

Muhammad ibn Ahmad ibn Muhammad Ibn Rushd was born on 14 April 1126 (520 AH) in Córdoba. His family was well known in the city for their public service, especially in the legal and religious fields. He was probably of Muladí and Berbernancestry. His grandfather Abu al-Walid Muhammad (d. 1126) was the chief judge (qadi) of Córdoba and the imam of the Great Mosque of Córdoba under the Almoravids. His father Abu al-Qasim Ahmad was not as celebrated as his grandfather, but was also chief judge until the Almoravids were replaced by the Almohads in 1146.

According to his traditional biographers, Ibn Rushd 's education was "excellent", beginning with studies in hadith (traditions of the Islamic prophet Muhammad), fiqh (jurisprudence), medicine and theology. He learned Maliki jurisprudence under al-Hafiz Abu Muhammad ibn Rizq and hadith with Ibn Bashkuwal, a student of his grandfather. His father also taught him about jurisprudence, including on Imam Malik's magnum opus the Muwatta, which Ibn Rushd went on to memorize. He studied medicine under Abu Jafar Jarim al-Tajail, who probably taught him philosophy too. He also knew the works of the philosopher Ibn Bajjah, and might have known him personally or been tutored by him. He joined a regular meeting of philosophers, physicians and poets in Seville which was attended by philosophers Ibn Tufayl and Ibn Zuhr as well as the future caliph Abu Yusuf Yaqub. He also studied the kalam theology of the Ashari school, which he criticized later in life.

By 1153 Ibn Rushd was in Marrakesh (present-day Morocco), the capital of the Almohad Caliphate, to perform astronomical observations and to support the Almohad project of building new colleges.

In 1169 Ibn Tufayl introduced Ibn Rushd to the Almohad caliph Abu Yaqub Yusuf. In a famous account reported by historian Abdelwahid al-Marrakushi the caliph asked Ibn Rushd whether the heavens had existed since eternity or had a beginning. Knowing this question was controversial and worried a wrong answer could put him in danger, Ibn Rushd did not answer. The caliph then elaborated the views of Plato, Aristotle and Muslim philosophers on the topic and discussed them with Ibn Tufayl. This display of knowledge put Ibn Rušd at ease; Ibn Rushd then explained his own views on the subject, which impressed the caliph. Ibn Rushd was similarly impressed by Abu Yaqub and later said the caliph had "a profuseness of learning I did not suspect".

After their introduction, Ibn Rushd remained in Abu Yaqub's favor until the caliph's death in 1184. When the caliph complained to Ibn Tufayl about the difficulty of understanding Aristotle's work, Ibn Tufayl recommended to the caliph that Ibn Rushd work on explaining it. This was the beginning of Ibn Rushd's massive commentaries on Aristotle; his first works on the subject were written in 1169.

In the same year, Ibn Rushd was appointed qadi (judge) in Seville. In 1171 he became qadi in his hometown of Córdoba. As qadi he would decide cases and give fatwas (legal opinions) based on the Islamic jurisprudence. The rate of his writing increased during this time despite other obligations and his travels within the Almohad empire. He also took the opportunity from his travels to conduct astronomical researches. Many of his works produced between 1169 and 1179 were dated in Seville rather than Córdoba. In 1179 he was again appointed qadi in Seville. In 1182 he succeeded his friend Ibn Tufayl as court physician and later the same year he was appointed the chief qadi of Córdoba, a prestigious office that had once been held by his grandfather.

In 1184 Caliph Abu Yaqub died and was succeeded by Abu Yusuf Yaqub. Initially, Ibn Rushd remained in royal favor but in 1195 his fortune reversed. Various charges were made against him and he was tried by a tribunal in Córdoba. The tribunal condemned his teachings, ordered the burning of his works and banished Ibn Rushd to nearby Lucena. Early biographers' reasons for this fall from grace include a possible insult to the caliph in his writings. However, modern scholars, who have little recourse to the original sources, attribute it to political reasons, which mostly represents their opinion. For example, the Encyclopedia of Islam speculates that the caliph distanced himself from Ibn Rushd to gain support from more orthodox ulema, who opposed Ibn Rushd and whose support al-Mansur needed for his war against Christian kingdoms. Such speculations and boldly expressing their opinions, without strong evidence, is common place by the Westerners when it comes to Muslim or Islamic matters.

After a few years, Ibn Rushd returned to court in Marrakesh and was again in the caliph's favor. He died shortly afterwards, on 11 December 1198 (9 Safar 595 in the Islamic calendar). He was initially buried in North Africa but his body was later moved to Córdoba for another funeral, at which future Sufi mystic and philosopher Ibn Arabi (1165–1240) was present.

48. Ibn al-Jawzi

Abd al-Raḥmān b. ʿAlī b. Muḥammad Abu 'l-Farash b. al-Jawzī, or briefly as Ibn al-Jawzī

(Arabic: ابن الجوزي),

(ca. 1116 – 16 June 1201),

Was jurisconsult, preacher, orator, heresiographer, traditionist, historian, judge, hagiographer, and philologist who played an instrumental role in propagating the Hanbali school of orthodox Sunni jurisprudence in his native Baghdad during the twelfth-century.

Contributions in Social Sciences

Ibn al-Jawzi is a prolific author in Islamic history. Al-Dhahabi states: "I have not known anyone amongst the 'ulama to have written as much as he (Ibn al-Jawzi) did". Some have suggested that he is the author of more than 700 works.

In addition to the topic of religion, Ibn al-Jawzi wrote about medicine as well. Like the medicinal works of Al-Suyuti, Ibn al-Jawzi's book was almost exclusively based on Prophetic medicine. Ibn al-Jawzi's work focused primarily on diet and natural remedies for both serious ailments such as rabies and smallpox and simple conditions such as headaches and nosebleeds.

Following are some of the works by Ibn al-Jawzi:

- A Collection of Great Fabricated Traditions (Arabic: الموضوعات الكبرى)
- Kitab Akhbar as-Sifat

- Sifatu al-Safwah, 5 parts, reworking of Hilyat al-Awliya' by Abu Nu'aym
- 'Ādāb al-Ḥasan al-Baṣrī wa-Zuhduh wa-Mawaʿiẓuh (آداب الحسن البصري وزهده ومواعظه): The Manners of al-Hasan al-Basri, his Asceticism, and his Exhortations
- Zad al-Masir fi Ilm al-Tafsir
- Talbīs Iblīs
- Tadhkirah Uli Al-Basāir fī Ma'rifah Al-Kabāir
- Gharīb Al-Ḥadīth
- Ahkam Al-Nisa
- Hifdh Al-'Umr
- Bahr Al-Damou'

Biographical Summary

Ibn al-Jawzi was born between 1113-19 CE to a "fairly wealthy family" in Baghdad, which "descended from Abu Bakr". His parents proceeded to give their son a "thorough education" in all the principal disciplines. Ibn al-Jawzi had the good fortune of studying under such notable scholars of the time as Ibn al-Zāghūnī (d. 1133), Abū Bakr al-Dīnawarī (d. 1137-8), Shaiykh Saiyed Razzaq Ali Gilani (d. 1208), Abū Manṣūr al-Jawālīkī (d. 1144-5), Abu 'l-Faḍl b. al-Nāṣir (d. 1155), Abū Ḥakīm al-Nahrawānī (d. 1161) and Abū Yaʿlā the Younger (d. 1163). Additionally, Ibn al-Jawzi began to be heavily influenced by the works of other scholars he read but whom he had never met personally, such as Abu Nu`aym (d. 1038), a Shafi'i Ashari mystic, al-Khatib al-Baghdadi (d. 1071), a Hanbali who had changed to Shafi'ism, and the prominent Hanbali thinker Ibn `Aqīl (d. ca. 1120), whom Ibn al-Jawzi would both praise and criticize in his later writings.

He was an adherent of the Ashari school of dialectical theology, an aspect of his thought that would later distinguish him from many of his fellow Hanbali thinkers. In his early works he criticized speculation in theology, in particular modernizing trends among the Sufis.

Ibn al-Jawzi began his career proper during the reign of al-Muqtafi (d. 1160), the thirty-first caliph of the Abbasid Caliphate, whose Hanbali vizier, Ibn Hubayra (d. 1165), served as a patron of Ibn al-Jawzi's scholarship. Beginning his scholarly career as a teaching assistant to his mentor Abū Ḥakīm al-Nahrawānī, who taught Hanbali jurisprudence in two separate schools, Ibn al-Jawzi succeeded al-Nahrawānī as "master of these two colleges" after the latter's death in 1161. A year or so prior to this, however, Ibn al-Jawzi had already begun his career as a preacher, as Ibn Hubayra had given him free rein to deliver his passionate sermons every Friday in the vizier's own house. After al-Muqtafi's death, the succeeding caliph, al-Mustanjid (d. 1170), called upon Ibn al-Jawzi to preach his sermons in the Caliph's Palace mosque – one of the most prominent houses of worship in the whole of Baghdad – during the three military interventions of the Fatimid Dynasty in the city. In these sermons, Ibn al-Jawzi is said to have "vigorously defended the prophetic precedent and criticized, not only all those whom he considered to be schismatics, but also the jurists who were too blindly attached to their own schools of law."

During the reign of the succeeding Abbasid caliph, al-Mustadi (d. 1180), Ibn al-Jawzi began to be recognized "as one of the most influential persons in Baghdad." As this particular ruler was especially partial to Hanbalism, Ibn al-Jawzi was given free rein to promote Hanbalism by way of his preaching throughout Baghdad. The numerous sermons Ibn al-Jawzi delivered from 1172 to 1173 cemented his reputation as the premier scholar in Baghdad at the time; indeed, the scholar soon began to be so appreciated for his gifts as an orator that al-Mustadi even went so far as to have a special dais (Arabic dikka) constructed specially for Ibn al-Jawzi in the Palace mosque. Ibn al-Jawzi's stature as a scholar continued to grow in the following years.

By 1179, Ibn al-Jawzi had written over one hundred and fifty works and was directing five colleges in Baghdad simultaneously. It was at this time that he told al-Mustadi to engrave an inscription onto the widely venerated tomb of Ibn Hanbal (d. 855) – the revered founder of the Hanbali rite – which referred to the famed jurist as "Imām." After the ascendancy of the new caliph, al-Nasir (d. 1235), to the Abbasid throne, Ibn al-Jawzi initially maintained amicable relations with the state power by way of his friendship with the caliph's Hanbali vizier, Ibn Yūnus (d. 1197). However, after the latter's dismissal and arrest – for unknown reasons – the caliph appointed as his successor the Shia Ibn al-Ḳaṣṣāb (d. ca. 1250). Although the reasons for the matter remain unclear in the historical record, al-Nasir eventually sentenced Ibn al-Jawzi to live under house arrest for five years. One of the possible reasons for this may be that Ibn al-Jawzi's relationship with the caliph had soured after the scholar had written a direct refutation of the ruler's policy in a particular matter. After five years in exile, Ibn al-Jawzi was eventually set free due to the pleading of al-Nasir's mother, whom the various chronicles describe as "a very devout woman" who pleaded with her son to free the famous scholar. Soon after his return to Baghdad, however, Ibn al-Jawzi died in 1201 AD, being seventy-four years old.

49. Ibn Jubayr

Ibn Jubayr

(Arabic: ابن جبير),

(1 September 1145 – 29 November 1217),

was a geographer, traveler and poet from al-Andalus.

Contributions in Social Sciences

Ibn Jubayr provides a highly-detailed and graphic description of the places he visited during his travels. The book differs from other contemporary accounts in not being a mere collection of toponyms and descriptions of monuments but containing observation of geographical details as well as cultural, religious and political matters. Particularly interesting are his notes about the declining faith of his fellow Muslims in Palermo after the recent Norman conquest and about what he perceived as the Muslim-influenced customs of King William II of Sicily under the Norman-Arab-Byzantine culture.

His writing is a foundation of the genre of work called Rihla, or the creative travelogue. It is a mix of personal narrative, description of the areas traveled and personal anecdotes.

Ibn Jubayr's travel chronicle served as a model for later authors, some of whom copied from it without attribution. Ibn Juzayy, who wrote the account of Ibn Battuta's travels in around 1355 AD, copied passages that had been written 170 years earlier by Ibn Jubayr that described Damascus, Mecca, Medina and other places in the Middle

East. Passages copied from Ibn Jubayr are also found in the writings of al-Sharishi, al-Abdari and Al-Maqrizi.

A surviving copy of Ibn Jubayr's manuscript is preserved in the collection of the Leiden University Library. The 210-page manuscript was produced in Mecca in 875 AH (1470 AD). The complete Arabic text was first printed in 1852 by the orientalist William Wright. An updated edition was printed in 1907 by Michael Jan de Goeje. A translation into Italian by Celestino Schiaparelli was published in 1906, a translation into English by Ronald Broadhurst was published in 1952, and a translation into French by Maurice Gaudefroy-Demombynes appeared in three volumes between 1949 and 1956.

Ibn Jubayr left Granada and crossed over the Strait of Gibraltar to Ceuta, then under Muslim rule. He boarded a Genoese ship on February 24, 1183 and set sail for Alexandria. His sea journey took him past the Balearic Islands and then across to the west coast of Sardinia. Offshore, he heard of the fate of 80 Muslim men, women and children who had been abducted from North Africa and were being sold into slavery. Between Sardinia and Sicily, the ship ran into a severe storm. He said of the Italians and Muslims on board who had experience of the sea that "all agreed that they had never in their lives seen such a tempest". After the storm, the ship went on past Sicily and Crete and turned south and crossed over to the North African coast. He arrived in Alexandria on March 26.

Saladin and the mamluks assured the protection of caravan routes that allowed travel to distant lands.

Everywhere that Ibn Jubayr traveled in Egypt, he was full of praise for the new Sunni ruler, Saladin. For example, he said, "There is no congregational or ordinary mosque, no mausoleum built over a grave, nor hospital, nor theological college, where the bounty of the Sultan does not extend to all who seek shelter or live in them". He pointed out that when the Nile did not flood enough, Saladin remitted the land tax from the farmers. He also said that "such is his (Salahuddin's) justice, and the safety he has brought to his high-roads that men in his lands can go about their affairs by night and from its darkness apprehend no awe that should deter them".

Ibn Jubayr, on the other hand, was very disparaging of the previous Shi'a dynasty of the Fatimids.

Of Cairo, Ibn Jubayr noted, the colleges and hostels that were erected for students and pious men of other lands by Saladin. In those colleges, students found lodging and tutors to teach them the sciences that they desired as well as also allowances to cover their needs. The care of the sultan also granted them baths, hospitals, and the appointment of doctors, who could even come to visit them at their place of stay who would be answerable for their cure. One of Saladin's other generous acts was that every day, 2000 loaves of bread were distributed to the poor. Also impressing Ibn Jubayr in the city was the number of mosques, estimated at between 8,000 and 12,000, with four or five of them often in the same street.

Upon arrival at Alexandria, Ibn Jubayr was angered by the customs officials who insisted on taking zakat from the pilgrims, regardless of whether or not they were obliged to pay. In the city, he visited the

Lighthouse of Alexandria, which was then still standing, and he was amazed by its size and splendor:

> One of the greatest wonders that we saw in this city was the lighthouse which Great and Glorious God had erected by the hands of those who were forced to such labor as 'Indeed in that are signs for those who discern'. Quran 15:75 and as a guide to voyagers, for without it they could not find the true course to Alexandria. It can be seen for more than seventy miles and is of great antiquity. It is most strongly built in all directions and competes with the skies in height. Description of it falls short, the eyes fail to comprehend it, and words are inadequate, so vast is the spectacle.

He was also impressed by the free colleges, hostels for foreign students, baths and hospitals in the city. They were paid for by awqaf and taxes.

He reached Cairo three days later. In the city, he visited the cemetery at al-Qarafah, which contained the graves of many important figures in the history of Islam. He noted that under Saladin, the walls of the citadel were being extended by the Mamluks with the object of reinforcing the entire city from any future Siege by Crusaders. Another work that he saw being built was a bridge over the Nile, which would be high enough not to be submerged in the annual flooding of the river. He saw a spacious free hospital, which was divided into three sections: for men, women and the insane. He saw the pyramids, and the Sphinx. He also saw a device that was used to measure the height of the Nile flood.

In Sicily, at the very late stages of his travels (December 1184 to January 1185), Ibn Jubayr recounted other experiences. He commented on the activity of the volcanoes:

> At the close of night, a red flame appeared, throwing up tongues into the air. It was the celebrated volcano (Stromboli). We were told that a fiery blast of great violence bursts out from air-holes in the two mountains and makes the fire. Often a great stone is cast up and thrown into the air by the force of the blast and prevented thereby from falling and settling at the bottom. This is one of the most remarkable of stories, and it is true.
>
> As for the great mountain in the island, known as the Jabal al-Nar [Mountain of Fire], it also presents a singular feature in that some years a fire pours from it in the manner of the `bursting of the dam'. It passes nothing it does not burn until, coming to the sea, it rides out on its surface and then subsides beneath it. Let us praise the Author of all things for His marvelous creations. There is no God but He.

Also striking Ibn Jubayr was the city of Palermo, which he described as follows:

> It is the metropolis of these islands, combining the benefits of wealth and splendor, and having all that you could wish of beauty, real or apparent, and all the needs of subsistence, mature and fresh. It is an ancient and elegant city, magnificent and gracious, and seductive to look upon. Proudly set between its open spaces and plains filled with gardens, with broad roads and avenues, it dazzles the eyes with its perfection. It is a wonderful place, built in the

Cordova style, entirely from cut stone known as kadhan [a soft limestone]. A river splits the town, and four springs gush in its suburbs.... The King roams through the gardens and courts for amusement and pleasure... The Christian women of this city follow the fashion of Muslim women, are fluent of speech, wrap their cloaks about them, and are veiled.

Ibn Jubayr also travelled to Medina, Mecca, Damascus, Mosul, Acre and Baghdad. At Basra, he saw how Indian timber was carefully used to make Lateen sail ships. He returned in 1185 by way of Sicily.

Biographical Summary

Ibn Jubayr was born in 1145 AD in Valencia, Spain. His family was from the Kinanah tribe. He was a descendant of 'Abdal-Salam ibn Jabayr, who, in 740 AD, had accompanied an army sent by the Caliph of Damascus to put down a Berber uprising in his Spanish provinces. Ibn Jubayr studied in the town of Xàtiva, where his father worked as a civil servant. He later became secretary to the Almohad governor of Granada.

Ibn Jubayr died on 29 November 1217 in Alexandria.

50. Al-Jawbari

ʿAbd al-Raḥīm ibn ʿUmar ibn Abī Bakr Jamāl al-Dīn al-Dimashqī, commonly known as al-Jawbarī

(Arabic: الجوبري),

(fl. 619/1222),

was a Syrian author and scholar known for his denunciation of alchemy.

Contributions in Social Sciences

Al-Jawbari wrote the "Book of Selected Disclosure of Secrets" (Kitāb al-mukhtār fī kashf al-asrār), exposing the fraudulence he had seen practiced by alchemists and money changers. He wrote of "the people of al-Kimya (alchemists) who know three hundred ways of making dupes." The book also describes the preparation of rose water.

Biographical Summary

Born in Jawbar, Syria, Al-Jawbari traveled extensively throughout the Islamic Empire, including visits as far as India. Among other locations, the scholar lived in Harrân and Kôniya.

51. Abd al-Latif al-Baghdadi

Muwaffaq al-Dīn Muḥammad ʿAbd al-Laṭīf ibn Yūsuf al-Baghdādī (Arabic: موفق الدين محمد عبد اللطيف بن يوسف البغدادي), (1162 Baghdad–1231 Baghdad), was a polymath; a physician, philosopher, historian, Arabic grammarian, Egyptologist, and traveler.

Contributions in Social Sciences

Abd al-Latif al-Baghdadi was a polymath, an expert in multiple disciplines. He was a medical scientist, archeologist, philosopher, and spiritualist.

ʿAbd-al-Laṭīf was well aware of the value of ancient monuments. He praised Muslim rulers for preserving and protecting pre-Islamic artefacts and monuments, but he also criticized them for failing to do this. He noted that the preservation of antiquities presented a number of benefits for Muslims.

- "monuments are useful historical evidence for chronologies";
- "they furnish evidence for Holy Scriptures, since the Qur'an mentions them and their people";
- "they are reminders of human endurance and fate";
- "they show, to a degree, the politics and history of ancestors, the richness of their sciences, and the genius of their thought".

While discussing the profession of treasure hunting, he notes that poorer treasure hunters were often sponsored by rich businessmen to go on archeological expeditions.

His manuscript was one of the earliest works on Egyptology. It contains a vivid description of a famine caused by the Nile failing to overflow its banks (which occurred during the author's residence in Egypt). Al-Baghdādī wrote that during the famine in Egypt in 1200 AD, he had the opportunity to observe and examine a large number of skeletons.

He wrote detailed descriptions on ancient Egyptian monuments.

Al-Baghdādī's manuscript was found in 1665 by the English orientalist Edward Pococke and is preserved in the Bodleian Library. Pococke published the Arabic manuscript in the 1680s. His son, Edward Pococke the Younger, translated the work into Latin, although he was only able to publish less than half of his work. Thomas Hunt attempted to publish Pococke's complete translation in 1746, although his attempt was unsuccessful. Pococke's complete Latin translation was eventually published by Joseph White of Oxford in 1800.

As far as philosophy is concerned, ʿAbd al-Laṭīf al-Baghdādī regarded philosophy to be in the service of religion. Therefore, he expected a true philosopher to have true insight, as well as a truly moral personality, thus verifying both belief and action. Apart from this he regarded the philosophers' ambitions as vain.

ʿAbd al-Laṭīf composed several philosophical works, among which is: Kitāb fī ʿilm mā baʿd al-ṭabīʿa an (book of metaphysics). The philosophical section of his Book "Kitāb al-Naṣīḥatayn" (two advices) is trend setting for the future work.

ʿAbd al-Laṭīf also penned two passionate pamphlets against the art of alchemy. Although he engaged in alchemy for a short while, he later abandoned the art completely by rejecting not only its practice, but also its theory.

Biographical Summary

Many details of ʿAbd al-Laṭīf al-Baghdādī's life are known from his biography as presented in Ibn Abī Uṣaybiʿah's literary history of medicine.

Abd al-Latif al-Baghdadi was born in Baghdad in 1162 AD. As a young man, he studied grammar, law, tradition, medicine, chemistry and philosophy. Avicenna (Ibn Sīnā) was his philosophical mentor. He travelled extensively and resided in Mosul (in 1189) where he studied the works of al-Suhrawardi before travelling on to Damascus (1190) and the camp of Saladin outside Acre (1191).

Here he met Baha al-Din ibn Shaddad and Imad al-Din al-Isfahani and acquired the Qadi al-Fadil's patronage. He went on to Cairo, where he met Abu'l-Qasim al-Shari'i, who introduced him to the works of al-Farabi.

In 1192 he met Saladin in Jerusalem and enjoyed his patronage, then went to Damascus again before returning to Cairo. He journeyed to Jerusalem and to Damascus in 1207-8, and eventually made his way via Aleppo to Erzindjan, where he remained at the court of the Mengujekid Ala'-al-Din Da'ud (Dāwūd Shāh). After the city was conquered by the Rūm Seljuk ruler Kayqubād II (Kayqubād Ibn Kaykhusraw), he returned to Baghdad in 1229, travelling back via Erzerum, Kamakh, Divriği and Malatya.

He died in Baghdad two years later in 1231 AD.

52. Ibn al-Athir

Ali 'Izz al-Din Ibn al-Athir al-Jazari

(Arabic: علي عز الدين بن الاثير الجزري),

(1160–1233),

was a historian and biographer.

Contributions in Social Sciences

Ibn Al-Athir lived a scholarly life in Mosul, often visited Baghdad and for a time traveled with Saladin's army in Syria. His chief work was a history of the world, al-Kamil fi at-Tarikh (The Complete History). Following are some of his works:

- Al-Kāmil fī al-tārīkh (الكامل في التاريخ): "The Complete History"; 11 volumes
- Al-Tārīkh al-bāhir fī al-Dawlah al-Atābakīyah bi-al-Mawṣil
- al-Usd al-ghābah fi ma'rifat al-ṣaḥābah: "The Lions of the Forest and the knowledge about the Companions"
- Al-Lubāb fī tahdhīb al-ansāb

Biographical Summary

He was from the Ibn Athir family. At the age of twenty-one he settled with his father in Mosul to continue his studies, where he devoted himself to the study of history and Islamic tradition.

Ibn al-Athir was born in 1160 AD, and belonged to the Shayban lineage of the large and influential Arab tribe Banu Bakr, who lived across upper Mesopotamia, and gave their name to the city of Diyar Bakir.

He was the brother of Majd ad-Dīn and Diyā' ad-Dīn.

Ibn Al-Athir lived a scholarly life in Mosul, often visited Baghdad and for a time traveled with Saladin's army in Syria. He later lived in Aleppo and Damascus. He died in the city of Mosul in 1233 AD.

53. Ibn Dihya al-Kalbi

Umar bin al-Hasan bin Ali bin Muhammad bin al-Jamil bin Farah bin Khalaf bin Qumis bin Mazlal bin Malal bin Badr bin Dihyah bin Farwah, better known as Ibn Dihya al-Kalbi

(Arabic: ابن دحية الكلبي),

(1152 AD in Valencia - 1235 AD in Cairo),

was a Moorish scholar of both the Arabic language and Islamic studies, and member of the Ahl al-Bayt.

Contributions in Social Sciences

In addition to his being renown as a linguistic scholar, Kalbi was also considered to be from the scholars of Prophetic narrations (Hadith), which was actually his primary focus of study, despite his fame primarily being for his knowledge of Arabic grammar and philology.

At the request by Al-Kamil, Ibn Dihya al-Kalbi wrote his most famous work: "al-Motrib min Ashaar Ahl al-Maghrib" "Arabic: المطرب من أشعار أهل المغرب" (Rhymes from the poetry of the people of the Maghreb) which was a collection of short biographies of the poets of Al-Andalus and Morocco, including their most famous poems. This work is valuable as it preserved some of the oral traditions of the time as well as details of the poets' lives.

He also wrote "al-Tanwir fi Mawlid al-Bashir al-Nadhir" a famous book on the birthday of the Prophet.

Biographical Summary

Ibn Diyha was born in 1152 AD in Valencia.

Kalbi initially pursued the study of Islamic Prophetic traditions (Hadith) in Spain, visiting every major city in the country and learning from their scholars and academics. In order to further his studies, he traveled from Al-Andalus to Marrakesh, Morocco.

Later, his study of Prophetic traditions took him through to Tunis in the year 590 AH, then through Africa and on to Egypt, the Levant, Baghdad, Erbil and Wasit in Iraq, and Isfahan and Nisapur in Greater Khorasan.

Kalbi was a student of Ibn Maḍā', chief judge of the Almohad Caliphate, and held immense respect for his teacher, who he referred to as "leader of all grammarians".

Ibn Diyha died in Cairo in 1235 AD

54. Al-Qifti

Jamāl al-Dīn Abū al-Ḥasan 'Alī ibn Yūsuf ibn Ibrāhīm ibn 'Abd al-Wahid al-Shaybānī

(Arabic: جمال الدين أبو الحسن علي بن يوسف بن إبراهي بن عبد الواحد الشيباني),

also called in Europe as 'Alī ibn Yūsuf al-Qifṭī (علي بن يوسف القفطي),

(1172–1248),

was an Egyptian historian, biographer-encyclopedist, patron, and administrator-scholar under the Ayyubid rulers of Aleppo.

Contributions in Social Sciences

His biographical dictionary Kitāb Ikhbār al-'Ulamā' bi Akhbār al-Ḥukamā (إخبار العلماء بأخبار الحكماء), tr. 'History of Learned Men', is an important source of Islamic biography.

Much of his vast literary output is lost, including his histories of the Seljuks, Buyids and the Maghreb, and biographical dictionaries of philosophers and philologists.

Al-Qifṭī wrote mainly historical works and of 26 recorded titles just two survive:

- Kitāb Ikhbār al-'Ulamā' bi Akhbār al-Ḥukamā (إخبار العلماء بأخبار الحكماء); abbrev. Ta'rikh al-Ḥukama (تاريخ الحكماء), 'The biographies and the books of the great philosophers'; a biographical dictionary of 414 physicians, philosophers and astronomers; the most important source of sciences in Islām.
- Inbā ar-Rawat 'alā 'Anbā an-Nuhat (3 vol.); synopsis (647/1249) by Muḥammad ibn 'Alī az-Zawanī.

Following are some of his lost works:

- Precious Pearls of the Account of the Master (Ad-Dur ath-Thamin fi 'Akhbar al-Mutīmīn) (الدر الثمين في أخبار المتيمين)
- Report of the Muhammad Poets, (Akhbar al-Muhammadin min al-Shuara), (posthumous); only fragments.
- History of Maḥmūd b. Sübüktigin (Sabuktakin) and His Sons'(wabanīhi, in al-Kubutī wabakīyat)
- History of the Seljuks, from the Beginning to the End of the Dynasty (Baqiat Tārīkh as-Siljūqīa) (بقية تاريخ السلجوقية)
- Apostles of Poets; arranged by al-Aba' up to Muḥammad bin Sa'īd; posthumous work written by al-Hasan ibn al-Haytham; History of the Poets; only poets named Muḥammad extant (Kitāb al-Muhmidīn min ash-Shu'ra'i; ratibah 'alā al-Ābā' wa balagh bīhī Muḥammad bin Sa'id.) (كتاب المحمدين من الشعراء. رتبه على الآباء وبلغ به محمد بن سعيد) (wa Katab 'an al-Hasan bin al-Haythm) (وكتب عن الحسن بن الهيثم)
- History of the Mirdasids (Akhbar al-Mirdas) (أخبار آلمرداس)
- The Biographies and Books of the Great Philosophers (Akhbar al-Alama bi Akhyar al-Hukama)(إخبار العلماء بأخيار الحكماء)
- Account of the Grammarians (Akhbar an-Nahwiyyin) (إخبار النحوين); survives only in abstract by Muh. b. Ahmad al-Dhahabi.
- Account of the Writers and their Writings (Akhbar al-Musanafin wa ma Sanafuh) (أخبار المصنفين وما صنفوه)
- History of the Yemen (Tarikh al-Yemen) (تاريخ اليمن)

- Egypt; in six parts ('Akhbār Misr, fi sitta 'Ajza') (أخبار مصر، في ستة أجزاء): including
  - History of Cairo until the reign of Salah al-Din; identical to Comprehensive Tarikh al-Qifti contained in the epitome of Ibn Maktum (d. 749/1348)
  - History of the Buyids
  - History of the Maghreb
  - Correction of Errors by al-Jawhari (Islāh Khilal as-Sahāhi, lil-Jawhrī) (إصلاح خلل الصحاح، للجوهري،)
  - Nahza al-Khater in Literature (Nahazat al-Khāṭr >> fi-l-Adab) (نهزة الخاطر» في الأدب); History of Scholarship (the Shaykhs of al-Kindi), a supplement to the Ansab of al-Baladhuri, etc.
  - Biographies of Ibn Rashiq, Abu Sa'id al-Sirafi.

Biographical Summary

'Alī al-Qifṭī, known as Ibn al-Qifṭī, was a native of Qift, Upper Egypt, the son of al-Qāḍī al-Ashraf, Yūsuf al-Qifṭī (b.548/1153), and the grandson of Ibrāhīm ibn 'Abd al-Wāḥid, al-Qāḍī al-Awḥad in the Ayyūbid court. Alī succeeded his father and grandfather into court administration but displayed scholarly inclinations. When the family left Qift in 1177, following the rising of a Fāṭimid Pretender, his father, Yūsuf, took up official posts in Upper Egypt and 'Alī completed his early education in Cairo.

In 583/1187 Yūsuf al-Qifṭī was appointed deputy to al-Qāḍī al-Fāḍil, chancellor and adviser to Ṣalāh al-Dīn at Jerusalem, and patron and benefactor of Maimonides. Al-Qifṭī spent many years studying and collecting material for his later works. When Ṣalāh al-Dīn died in 598/1201 and his brother, Malik al-'Ādil, usurped his nephew's position in Jerusalem, Ibn al-Qifṭī's father fled to Ḥarran into the service of Ṣalāh al-Dīn's son Ashraf. Ibn al-Qifṭī sought patronage in Aleppo as secretary to the former governor of Jerusalem and Nablus, Fāris al-Din Maimūn al Qaṣrī, the then vizier to the Ayyubid emir Ṣalāh al-Dīn's third son, Malik aẓ-Ẓāhir Ghāzi.

He was recognized as an effective administrator of the fiefs and when the vizier died in 610/1214 aẓ-Ẓāhir appointed him "khāzin", or Dīwān of Finance, despite his own preference for study. On aẓ-Ẓāhir's death in 613/1216 al-Qifti retired but was re-appointed in 633/1236 by aẓ-Ẓāhir's successor.

According to his protégé and biographer, Yaqūt, al-Qifti already held the honorific title "al-Qāḍī 'al-Akram al-Wazir" (most noble judge chief minister). After a five-year sabbatical al-Qifṭī again resumed the office and held it up to his death in 646/1248.

Throughout his life al-Qifṭī advocated scholarship and sought to pursue a literary career despite heavy constraints of high office.

55. Ibn al-Abbar

Hāfiẓ Abū Abd Allāh Muḥammad ibn 'Abdullah ibn Abū Bakr al-Qudā'ī al-Balansī

(Arabic: أبو عبد الله محمد بن عبد الله بن أبي بكر بن عبد الله بن عبد الرحمن القضاعي البلنسي),

(also known as Ibn al-Abbār),

(1199–1260),

was a historian, a jurist, a hadith scholar, and a diplomat.

Contributions in Social Sciences

Of the forty-five books by Ibn al-Abbār eight survived.

Kitāb al-Takmila li Kitāb al-ṣila (كتاب التكملة لالكتاب الصلة) is at-Takmila ('Supplement') to the Ta'rīkh' Ulamā 'al-Andalus ('History of the scholars of Andalusia'). The history was originally written by Ibn al-Faradi (962-1013). Ibn Bashkuwāl (1101-1183) had written a sequel history titled Ṣila fī ta'rīkh a'immat al-Andalus. Ibn al-Abbār's supplement is to "Ṣila fī ta'rīkh a'immat al-Andalus".

Ibn al-Abbār's Valencian master Abū al-Rabi 'ibn al-Sālim prompted him to complete the two works of the previous generation. Ibn al-Abbār began working on "The Supplement" in 1233 at Valencia, and finally completed it in Tunis. It lists (in alphabetical order) more than three thousand scholars in the literary and cultural history of Muslim Spain. In the introduction, the author makes clear his concern about the threats to his homeland and his desire to save some of the intellectual heritage for posterity.

This work by Ibn al-Abbār has been published in several incomplete editions from different manuscripts.

- Takmila from Fez MS, ed., Alfred Bel and Mohamed Bencheneb (Algiers, 1920); 652 biographies of the first five letters from the letter alif.
- Takmila from Cairo MS, ed., 'Abd al-'Aṭṭār al-Ḥusayni (2 vols., Baghdad and Cairo, 1956), begins with the letter alif, comprising 2188 biographies.

Following are some editions of translations of this work of Ibn al-Abbār:

- Complementum libri assilah: dictionarium biographicum (in Arabic; Romero Matriti, 1877); vols., 5&6, vols., 7&8, vols., 9&10.
- Complementum Libri as-Sila, ed., Francisco Codera Zaidin, Madrid, Biblioteca Arabo-Hispana, 2 vols., nos. V-VI, 1888–89); 2152. biographies from the Escurial copy, and 600 from Algiers MS; begins with the letter ج(ğīm).
- Miscelaneas de estudios y textos árabes, eds., Maximiliano Alarcón and Cándido Ángel González Palencia, Madrid, 1915, pp. 147–690); an appendix to previous, from a manuscript of Cairo, with biographies nos., 2150 - 2892.

Following are some additional works by Ibn al-Abbār:

- Kitāb al-ḥulla al-siyarā ('Book of the embroidered tunic'), finished at Béjaïa in 1248/49, compendium of the poetic-literary field.

- Tuḥfat al-qādim (تحفة القادم), 'Newcomer's gift'; life and works of the Andalusian poets of his time.
- I'tāb al-kuttāb, a short collection of stories of disgraced and rehabilitated officials, written during his exile at Béjaïa.
- Al-'Arba'ūn (الأربعون), 'The Forty Ahadith.
- Durar al-simṭ fī khabar al-sibṭ (درر السِّمط في خبر السِّبط), 'Pearl necklace on the reports of the Israelites'; written during his second stay at Béjaïa, a religious work of Shiite leanings defending the persecuted lineage of 'Ali.
- Dīwān ('collected poems') of Ibn al-Abbār.
- Ya'nī al-Ḥusayn ibn 'Alī (يعني الحسين بن علي) 'meaning Husayn ibn Ali'

Following are some references to present day research on التكملة لاالكتاب الصلة by Ibn al-Abbār:

- Bel, Alfred; Bencheneb, Mohamed (1918). "The preface of Ibn al-Abbar to his" Takmila as-sila". African Review. Paris (294): 306–335.
- Ghedira, Ameur (1957). "An unpublished treatise of Ibn al-Abbar with Shia tendency". Al-Andalus. Madrid- Granada. 22: 30–54.
- Khallikān (Ibn), Aḥmad ibn Muḥammad (1843). Ibn Khallikan's Biographical Dictionary (tr. Wafayāt al-A'yān wa-al-Anbā Abnā' al-Zamān). Vol. ii. Translated by McGuckin de Slane, William. London: W.H. Allen. p. 424, n.3.

- Meouak, Mohamed (1985). "Ibn al-Abbār's "Takmilla": notes and observations about his editions". Journal of the Muslim West and the Mediterranean. 40: 143–146.
- "Ibn al-Abbar, politic i escriptor derab valencià (1199-1260)". proceedings of the international congress "Ibn al-Abbar i el seu temps (in Spanish). Valencia: Generalitat Valenciana. 1990.
- Ibn al-Abbar, politic i escriptor Arab valencia (1199–1260): Actes del Congres Internacional "Ibn Al-Abbar i el seu temps," Onda, 20-22 febrer, 1989 by Mikel Epalza, Jesus Huguet (review Journal of the American Oriental Society, Vol. 112, No. 2 (Apr. - Jun., 1992), pp. 313–314)

Biographical Summary

Ibn al-Abbār was born in 1199 in Valencia, Andalusia. His family were of Yemeni Arab ("al-Qudā'ī") ancestry, but had lived for generations in the village of Onda. As an only son, his father, a scholar, a faqīh (jurist), and a poet, gave him the best education. He was taught by famous scholars of the time, such as Abū l-Rabi 'ibn al-Sālim, and cultivated in jurisprudence and poetry.

He travelled through al-Andalus. In 1222, while in Badajoz, he learned of his father's death. He returned to Valencia.

Ibn al-Abbār became secretary (kātib) to the governor Abū Zayd. In 1229 a revolt against the Almohades forced Abū Zayd to flee the city. Ibn al-Abbār accompanied him when Abū Zayd took refuge with king James I of Aragon. When Abū Zayd converted to Christianity, Ibn al-Abbār abandoned him and returned to Valencia in 1231 to become vizier to the new ruler, Abū Jamil ibn Zayyan ibn Mardanish, whom he knew from an earlier period.

Around 1235, Ibn al-Abbār was qadi (judge) for a time at Dénia. In 1236 Cordova fell to Ferdinand III of Castile and in 1237 James I of Aragon defeated Ibn Mardanish at the Battle of the Puig. The siege of Valencia began soon after. Abu Jamil sent Ibn al-Abbār to seek help from Abū Zakariyā Yaḥyā, the Hafsid sultan of Tunis. The ambassador declaimed before the Sultan a famous "qasīda" celebrating "al-Andalus" and deploring its tragic situation. Abū Zakariyā sent a fleet of twelve ships, which failed however to reach the blockaded port of Valencia, and was forced to anchor at Dénia. Subsequently Ibn al-Abbār was charged by the emir with negotiating the surrender of Valencia, which was signed on September 29, 1238. The two fled to Dénia and Murcia, and in 1240 Ibn al-Abbār emigrated permanently to Tunis.

He was once again welcomed by Abū Zakariyā, and appointed head of his chancery and his panegyrist. With enemies at court (notably the vizier Ibn Abul Husayn), he was replaced and exiled to Béjaïa in 1248. Abū Zakariyā before his death in 1249 had forgiven and recalled ibn al-Abbār, and he became counsellor to Abū Zakariyā's successor, Muhammad I al-Mustansir. However, ibn al-Abbār was again banished to Bejaia in 1252. After the fall of the Abbāsid Caliphate of Baghdad (1258), Muhammad I al-Mustansir had proclaimed himself caliph (and was recognized as such in Mecca and Medina). In 1259, Ibn al-Abbār was again forgiven and recalled to Tunis. Soon after he was arrested, and sentenced to be burnt at the stake. He was put to death by order of al-Mustansir, the ruler of Tunis, on the 6th of January 1260, and his body along with his books were burned.

An account of this is given by Ibn Khaldūn in his History of the Berbers (Kitāb al-'Ibar).

56. Ibn al-Adim

Kamāl al-Dīn Abu ʾl-Ḳāsim ʿUmar ibn Aḥmad ibn Hibat Allāh Ibn al-ʿAdīm

(Arabic: كمال الدين عمر بن أحمد ابن العديم),

(1192–1262),

was a biographer and historian from Aleppo.

Contributions in Social Sciences

Ibn al-ʿAdīm is best known for his work: Bughyat al-Talab fī Tārīkh Ḥalab (بغية الطلب في تاريخ حلب; Everything Desirable about the History of Aleppo). It is a multi-volume collection of biographies of famous men from Aleppo, introduced with a volume on the geography and traditions of the region.

It is saved in part in manuscripts in the library of sultan Ahmed III in Topkapi Palace.

Ibn al-ʿAdīm also published a chronicle version of the work titled Zubdat al-Halab fi ta'arikh Halab (زبدة الحلب في تأريخ حلب; The Cream of the History of Aleppo). A copy of this work reached the library of Jean-Baptiste Colbert and then the Bibliothèque nationale de France.

Selections of this work were published with Latin translation by Georg Freytag in 1819.

Ibn al-ʿAdīm's historical sources are various, some oral and some written, and two of the more famous are Usama ibn Munqidh and Ibn al-Qalanisi.

Another work of Ibn al-ʿAdīm is a guide for making perfumes titled Kitab al-Wuslat (or Wasilat) ila-l- Habib fi Wasf al-Tayibat wal-tibb (Houtsma 1927).

Ibn al-ʿAdīm is an important source of knowledge on the Syrian Assassins, first analyzed by Silvester de Sacy (Lewis 1952).

Biographical Summary

Ibn al-ʿAdīm was born in Alippo in 1192 AD, and he died in 1262 in Egypt.

Numerous Ayyubid rulers entrusted Ibn al-Adim as a diplomatic ambassador. On his last mission in 1260, he was sent to Egypt seeking military assistance against the Mongols.

57. Ibn Abi Usaibia

Ibn Abī Uṣaybiʿa Muʾaffaq al-Dīn Abū al-ʿAbbās Aḥmad Ibn Al-Qāsim Ibn Khalīfa al-Khazrajī

(Arabic: ابن أبي أصيبعة), commonly referred to as Ibn Abi Usaibia, (1203–1270),

was a physician from Syria. He compiled a biographical encyclopedia of notable physicians, up to the year 1252 AD.

Contributions in Social Sciences

Ibn Abi Usaibia wrote an encyclopedia of physicians under the title, Uyūn ul-Anbāʾ fī Ṭabaqāt al-Aṭibbā (Arabic: عيون الأنباء في طبقات الأطباء). The title can be translatable loosely and expansively as "Sources of News on Classes of Physicians", though commonly it is translated into English as History of Physicians, Lives of the Physicians, Classes of Physicians, or Biographical Encyclopedia of Physicians. The main focus of the book's 700 pages is physicians of Islamic era. A first version appeared in 1245–1246 and was dedicated to the Ayyubid physician and vizier Amīn al-Dawlah. A second and enlarged recension of the work was produced in the last years of the life of the author, and circulated in at least two different versions, as shown by the extant manuscripts.

The text has been published in five editions. When the first edition by August Müller (Cairo, 1882), appeared under the pseudonym "Imrū l-Qays", it was found to be marred by typos and errors; a corrected version was subsequently issued (Königsberg, 1884). Relying on

Müller's work, Niẓār Riḍā published a non-critical edition of the text in Beirut in 1965, which was subsequently reworked by Q̱āsim Wahhāb for yet another edition issued in Beirut in 1997. ʿĀmir al-Najjār published his own critical edition (not based on Müller) in Cairo in 1996.

A team of academicians from the universities of Oxford and Warwick has published a new critical edition and a full annotated English translation of the Uyūn al-Anbā. Their work is available in Open Access at Brill Scholarly Editions.

In 2020, a new translation was published by Oxford World's Classics under the name Anecdotes and Antidotes: A Medieval Arabic History of Physicians.

Biographical Summary

Ibn Abi Usaibia was born in 1203 AD at Damascus, a member of the Banu Khazraj tribe. The son of a physician, he studied medicine at Damascus and Cairo. In 1236, he was appointed physician to a new hospital in Cairo, but the following year he took up an offer by ruler of Damascus, of a post in Salkhad, near Damascus, where he lived until his death. His only surviving work is Lives of the Physicians. In that work, he mentions another of his works, but it has not survived.

Ibn Abi Usaibia died in 1270 AD.

58. Ibn Malik

Abu 'Abd Allah Jamal al-Din Muḥammad ibn Abd Allāh ibn Malik al-Ta'i al-Jayyani

(Arabic: ابو عبدالله جمال الدين محمد بن عبدالله بن عبدالله بن مالك الطائي الجياني النحوي),

(1204– 21 February 1274),

was a grammarian born in Jaén.

Contributions in Social Sciences

He was a senior master at the Adiliyya Madrasa. His reputation in Arabic literature was cemented by his treatise titled al-Khulāsa al-alfiyya (known also as simply Alfiya). It is a versification of Arabic grammar, for which at least 43 commentaries have been written.

Biographical Summary

Ibn Malik was born in Jaén, Andalusia. He was either born in 1203/4 AD or 1204/5 AD.

Ibn Malik left al-Andalus for the Near East, where he worked on Arabic language and literature in Aleppo and Hamāt. He eventually settled in Damascus, where he began the most productive period of his life.

Ibn Malik died February 21, 1274 AD.

59. Zakariya al-Qazwini

Abū Yaḥyā Zakariyyā' ibn Muḥammad ibn Maḥmūd al-Qazwīnī
(Arabic: أبو يحيى زكرياء بن محمد بن محمود القزويني),
(born c. 1203 in Qazvin (Iran) and died 1283),
was a Persian cosmographer and geographer of Arab ancestry.

Contributions in Social Sciences

His most famous work is the 'Ajā'ib al-makhlūqāt wa-gharā'ib al-mawjūdāt (lit. 'Wonders of the Creation and Unique [phenomena] of the Existence'), a seminal work in cosmography.

This treatise, frequently illustrated, was popular and is preserved today in many copies. The first edition was written around 659H/1260-61 in Persian Language. During 678H/1279-80 it was edited and translated into Arabic language, and later also into Turkish.

Al-Qazwini also authored the geographical dictionary titled Āthār al-bilād wa-akhbār al-'ibād (lit. 'Monuments of the Lands and Historical Traditions about Their Peoples').

Both of these treatises reflect extensive reading and learning in a wide range of disciplines.

Biographical Summary

Zakariyyā' al-Qazwīnī belonged to a family of jurists originally descended from Anas bin Malik (a companion of the prophet). The family had been well established in Qazvin long before al-Qazwini was born.

Born in Qazvin, Iran, al-Qazwini served as a legal expert and judge in several localities in Iran. He traveled around in Mesopotamia and the Levant, and finally was patronized by the Ilkhanid governor of Baghdad, Ata-Malik Juvayni (d. 1283 CE). Al-Qazwini dedicated his famous cosmography titled ʿAjāʾib al-makhlūqāt wa-gharāʾib al-mawjūdāt (lit. 'Wonders of the Creation and Unique [phenomena] of the Existence') to the governor.

Zakariyyāʾ al-Qazwīnī died in 1283 AD.

60. Ibn Sa'id al-Maghribi

Abū al-Ḥasan ʿAlī ibn Mūsā ibn Saʿīd al-Maghribī

(Arabic: علي بن موسى المغربي بن سعيد),

also known as Ibn Saʿīd al-Andalusī,

(1213–1286),

was a geographer, historian, poet, and an important collector of poetry from al-Andalus in the 12th and 13th centuries.

Contributions in Social Sciences

Ibn Said al-Maghribi wrote or compiled at least forty works on various branches of knowledge.

Ibn Said's best-known achievement was the completion of the fifteen-volume al-Mughrib fī ḥulā l-Maghrib ('The Extraordinary Book on the Adornments of the West'), which had been started over a century before by Abū Muḥammad al-Ḥijārī (1106–55) at the behest of Ibn Said al-Ḥijārī's great-grandfather, 'Abd al-Malik.

Abū Muḥammad al-Ḥijārī had completed 6 volumes.

'Abd al-Malik added to them two of 'Abd al-Malik's sons (Ibn Said's grandfather and great uncle).

Ibn Said al-Ḥijārī's father worked on it further; and Ibn Said completed it.

The work is also known as the Kitāb al-Mughrib ('book of the Maghrib'). It is an anthology of poetry, collecting information on the poets of Maghreb organized by geographical origin.

Part of the Kitāb al-Mughrib circulated separately as Rāyāt al-mubarrizīn wa-ghāyāt al-mumayyazīn (Banners of the Champions and the Standards of the Distinguished), which Ibn Said compiled in Cairo, completing it on 21 June 1243.

As an indefatigable traveler, Ibn Said was profoundly interested in geography. In 1250 he wrote his Kitab bast al-ard fi 't -t ul wa-'l-'ard (The Book of the Extension of the Land on Longitudes and Latitudes). His Kitab al-Jughrafiya (Geography) embodies the experience of his extensive travels through the Muslim world and on the shores of the Indian Ocean. He also gives an account of parts of northern Europe including Ireland and Iceland. He visited Armenia and was at the Court of Hulagu Khan from 1256 to 1265.

Another one of Ibn Said's works include Al-Ṭāliʿ al-Saʿīd fī Tārīkh Banī Saʿīd, a history of the Banū Saʿīd. It is preserved only fragmentarily, in quotation by others.

Biographical Summary

Ibn Said was born in 1213 AD at Alcalá la Real near Granada to a prominent family which was descended from Ammar ibn Yasir, who was a Companion of the Prophet. He grew up in Marrakesh. Many of his family members were literary figures. He subsequently studied in Seville and stayed in Tunis, Alexandria, Cairo, Jerusalem and Aleppo. At the age of 30, he undertook a pilgrimage to Mecca. He was also a close friend of the Muwallad poet, Ibn Mokond Al-Lishboni (of Lisbon). His last years were spent in Tunis, and he died there in 1286.

61. Ibn Manzur

Muhammad ibn Mukarram ibn Alī ibn Ahmad ibn Manzūr al-Ansārī al-Ifrīqī al-Misrī al-Khazrajī

(Arabic: محمد بن مكرم بن علي بن أحمد بن منظور الأنصاري الإفريقي المصري الخزرجي),

also known as Ibn Manẓūr (Arabic: إبن منظور),

(June–July 1233 – December 1311/January 1312),

was a lexicographer of the Arabic language and author of a large dictionary, Lisan al-ʿArab (لسان العرب; lit. 'The Tongue of the Arabs'.

Contributions in Social Sciences

Lisān al-ʿArab (لسان العرب, "Tongue of Arabs") was completed by Ibn Manzur in 1290. Occupying 20 printed book volumes (in the most frequently cited edition), it is the best-known dictionary of the Arabic language, as well as one of the most comprehensive.

Ibn Manzur compiled it from other sources to a degree. The most important sources for it were the Tahdhīb al-Lugha of Azharī, Al-Muḥkam of Ibn Sidah, Al-Nihāya of Ibn Athīr and Jauhari's Ṣiḥāḥ, as well as the ḥawāshī (glosses) of the latter (Kitāb at-Tanbīh wa-l-Īḍāḥ) by Ibn Barrī.

It follows the Ṣiḥāḥ in the arrangement of the roots: The headwords are not arranged by the alphabetical order of the radicals as usually done today in the study of Semitic languages, but according to the last radical - which makes finding rhyming endings significantly easier. Furthermore, the Lisān al-Arab notes its direct sources, but not or seldom their sources, making it hard to trace the linguistic history of certain words. Murtaḍá al-Zabīdī corrected this in his Tāj al-ʿArūs, that itself goes back to the Lisān.

The Lisān, was already printed in the 18th century in Istanbul.

Following are published editions of the Lisān al-'Arab:

- Bullag Misr al-Matb'ah al-Kubra al-'Amiriyah Egypt; 1883, vol., 1
- Al-Maṭbaʿa al-Kubra al-Amirīya, Bulaq; 1883 - 1890, vols., 20
- Dar Sadir, Beirut; 1955 - 1956, vols., 15.
- Ādāb al-Ḥawza, Iran; 1984, vols., 18

Ibn Manzur dedicated most of his life to excerpts from works of historical philology. He is said to have left 500 volumes of this work.

Following are some other works by Ibn Manẓūr:

- Aḫbār Abī Nuwās, a bio-bibliography of the Arabic-Persian poet Abu Nuwas; printed (with commentary by Muhammad Abd ar-Rasul) 1924 in Cairo as well as published by Shukri M. Ahmad 1952 in Baghdad.
- Muḫtaṣar taʾrīḫ madīnat Dimašq l-Ibn ʿAsākir, summary of the history of Damascus by Ibn 'Asakir.
- Muḫtaṣar taʾrīḫ madīnat Baġdād li-s-Samʿānī, summary of the history of Baghdad by al-Samʿānī (d. December 1166).
- Muḫtaṣar Ǧāmiʿ al-Mufradāt, summary of the treatise about remedies and edibles by al-Baiṭār.
- Muḫtār al-aġānī fi-l-aḫbār wa-t-tahānī, a selection of songs; printed 1927 in Cairo.

- Niṯār al-azhār fī l-layl wa-l-nahār, a short treatise on astronomy about day and night as well as the stars and zodiacs; printed 1880 in Istanbul.
- Taḏkirāt al-Labīb wa-nuzhat al-adīb (if following Fück identical with Muḥammad b. Mukarram), served al-Qalqaschandi as a source.

Biographical Summary

Ibn Manzur was born in 1233 in Ifriqiya (present day Tunisia). He was of North African Arab descent, from the Banu Khazraj tribe of Ansar as his nisba al-Ansārī al-Ifrīqī al-Misrī al-Khazrajī suggests. Ibn Hajar reports that he was a judge (qadi) in Tripoli, Libya and Egypt and spent his life among Kuttāb al-Inshā᾿ in the Diwan al-Insha', an office that was responsible among other things for correspondence, archiving and copying.

He died around the turn of the years 1311/1312 in Cairo.

62. Al-Dimashqi

Shams al-Din al-Ansari al-Dimashqi or simply al-Dimashqi

(Arabic: شمس الدين الأنصاري الدمشقي),

(1256–1327),

was a geographer.

Contributions in Social Sciences

Born in Damascus he mostly wrote of his native land, the Greater Syria (Bilad ash-Sham). He completed his main work in 1300, the time of the complete withdrawal of the Crusaders. He is a contemporary of the Mamluk sultan Baibars, the general who led the Muslims in war against the Crusaders. His work is especially of value in as the Crusader Chronicles.

Al-Dimashqi also gives detailed accounts of the islands in Maritime Southeast Asia, its inhabitants, flora, fauna and customs.

He mentions:

"the country of Champa...is inhabited by Muslims and idolaters. Islam arrived there during the time of Caliph Uthman...and Ali. Many Muslims who were expelled by the Umayyads and by Al-Hajjaj, fled there, and since then a majority of the Cham have embraced Islam."

Of their rivals the Khmer, Al-Dimashqi mentions:

they inhabit the island of Komor (Khmer), also called Malay Island, are many towns and cities, rich-dense forests with huge, tall trees, and white elephants; they supplemented their income from the trade routes not only by exporting ivory and aloe, but also by engaging in piracy and raiding on Muslim and Chinese shipping.

Al-Dimashqi's writings on Syria were published in St. Petersburg in 1866 by M.A.F Mehren, and this edition was later used for the English translation by Guy Le Strange in 1890.

Biographical Summary

Al-Dimashqi was born in Damascus in 1256 AD, and he died while in Safad, in 1327.

63. Al-Nuwayri

Shihāb al-Dīn Ahmad bin 'Abd al-Wahhāb al-Nuwayri

(Arabic: شهاب الدين أحمد بن عبد الوهاب النويري),

(born April 5, 1279 in Akhmim, present-day Egypt – died June 5, 1333 in Cairo),

was an historian from Egypt, and a civil servant of the Bahri Mamluk dynasty.

Contributions in Social Sciences

Al-Nuwayri is notable for his compilation of a 9,000-page encyclopedia of the Mamluk era, titled The Ultimate Ambition in the Arts of Erudition (نهاية الأرب في فنون الأدب, Nihāyat al-arab fī funūn al-adab), which pertained to zoology, anatomy, history, chronology, amongst others. Al-Nuwayri started his encyclopedia around the year 1314 and completed it in 1333.

Al-Nuwayri's encyclopedia, The Ultimate Ambition in the Arts of Erudition, was divided into five sections (books):

- Geography and astronomy
- Man, and what relates to him
- Animals
- Plants
- History

The first four subjects comprised 10 volumes, while the last filled 21 volumes.

Al-Nuwayri based his encyclopedia on several earlier works. Some wholly original researches are in the discussion of financial secretaryship in book two, and some of the historical material in book five. Much of the work was a compilation of a number of texts including Delightful Concepts and the Path to Precepts (Mabahij al-fikar wa manahij al-'ibar) by Jamal al-Din al-Watwat, and Avicenna's Canon of Medicine.

Al-Nuwayri is also known for his extensive work regarding the Mongols' conquest of Syria.

Biographical Summary

The name Al-Nuwayri is a nisba referring to the village of Al-Nuwayra. Al-Nuwayri was born April 5, 1279 in Akhmim, Egypt. For most of his childhood, he lived in Qus in Upper Egypt, where he studied with Ibn Daqiq al-'Id. He later studied at Al-Azhar University in Cairo, specializing in the study of the hadith and the sira, in addition to history. Skilled in calligraphy, he reportedly made a copy of Sahih al-Bukhari which he sold for 1000 dinars. He worked as a civil servant in the administration of Sultan An-Nasir Muhammad starting at age 23, serving in various roles including property manager for the Sultan and superintendent of army finances in Tripoli. At some point after 1312, he retired from government service and took a job copying manuscripts in order to support himself while compiling his encyclopedia. He died on June 5, 1333 in Cairo.

64. Fath al-Din ibn Sayyid al-Nas

Muhammad bin Muhammad al-Ya'mari, better known as Fatḥ al-Dīn Ibn Sayyid al-Nās, specialized in the field of Hadith. He was well known for his biography of the Prophet.

Contributions in Social Sciences

Ibn Sayyid al-Nas' biography of the prophet Muhammad is well known. Some of the isnads, or chains of narration establishing the historicity of claims are unique. His isnad are well researched.

Ibn Sayyid al-Nas' work uniquely provides isnad in support of the events in the prophetic biography. For example, isnad for some of the narrations in the sira by Ibn Hisham are available only in the research of Ibn Sayyid al-Nas.

During his time, Ibn Sayyid al-Nas' was considered one of Cairo's greatest composers of poetry in praise of the Prophet. Ibn Sayyid al-Nas along with Abu Hayyan al-Gharnati often presided as judges over the poetic contests during the reign of Mamluk sultan Al-Nasir Muhammad.

Slimane of Morocco, the sultan of Morocco in the early 1800s, who greatly restricted the acceptable reading material in his sultanate, designated Ibn Sayyid al-Nas's prophetic biography as one of only two approved works.

Ibn Sayyid al-Nas was respected among hadith circles for his transmissions of a recension of Sahih al-Bukhari. In regard to the widely reported raid of Hudhayl, Ibn Sayyid al-Nas' transmission is nearly identical to the narrations of Muhammad al-Bukhari himself, save seven small differences, six copyist errors and one difference in a single word.

Biographical Summary

Ibn Sayyid al-Nas was born in 1272 AD in Egypt. He was descended from an Andalusian family from Seville. The family fled due to hostility from Christians, who eventually took the city in 1248.

His grandfather Abu Bakr Muhammad bin Ahmad was born in 1200 and settled in Tunis, where Ibn Sayyid al-Nas' father was born in October 1247. His grandfather died in 1261.

Ibn Sayyid al-Nas died in the year 1334. He was known as an adherent of the Zahiri school of Sunni Islam.

65. Hamdallah Mustawfi

Hamdallah Mustawfi Qazvini

(Persian: حمدالله مستوفی قزوینی),

(1281 – after 1339/40),

was a Persian official, historian, geographer and poet. He lived during the last era of the Mongol Ilkhanate, and the interregnum that followed.

Contributions in Social Sciences

Mustawfi's first work was the Tarikh-i guzida ("Excerpt History"), a world history, narrating the events of the prophets, the pre-Islamic kings of Iran, and the Islamic world; based on the then incomplete Zafarnamah. Hamdallah Mustawfi credits, in a thorough manner, the previous sources that he used.

Tarikh-i guzida contains important information after the death of the Ilkhanate monarch Ghazan in 1304. The political tale concludes in a positive tone, with Ghiyath al-Din Muhammad being appointed to vizierate of the Ilkhanate. The penultimate chapter contains the lives of distinguished scholars and poets, whilst the last describes Qazvin and gives a reportage of its history.

Mustawfi's second work was the Zafarnamah ("Book of Victory"), a continuation of Ferdowsi's Shahnameh ("Book of Kings"). He completed the work in 1334, consisting of 75,000 verses, reporting the history of the Islamic era up until the Ilkhanate era. Zafarnamah is a unique primary source for the reign of the Ilkhanate monarch Öljaitü

(r. 1304–1316) and that of his successor, Abu Sa'id Bahadur Khan (r. 1316–1335). The importance of the work was acknowledged by the Timurid-era historian Hafiz-i Abru, who incorporated much of it in his Dhayl-e Jame al-tawarikh.

Like the Tarikh-i guzida, the Zafarnamah has a positive conclusion, with Abu Sai'd Bahadur Khan successfully quelling a revolt, followed by peace. Mustawfi later composed a prose continuation of the Zafarnamah, which mentions Abu Sai'd Bahar Khan's death and the turmoil that followed in Iran.

Mustawfi's most prominent work is the Nuzhat al-qulub ("Hearts' Bliss"), which is virtually the only source to describe the geography and affairs of the Ilkhanate era. The source gives vital information about the government, commerce, economic life, sectarian conflicts, tax-collection and other similar topics. Just like his Tarikh-i guzida and Zafarnamah, Mustawfi rejects to have expertise in the field, and states that he was encouraged by his friends to write the work. He also thought that an available source in Persian would be helpful, due to most geographical sources about Iran being in Arabic (such as the works of Abu Zayd al-Balkhi and Ibn Khordadbeh).

The work is also considered a substantial contribution to the ethno-national history of Iran. Mustawfi notably uses the term "Iran" in his work. Since the fall of the Iranian Sasanian Empire in 651, the idea of Iran or Iranzamin ("the land of Iran") as a political entity had disappeared. However, it did remain as an element of the national sentiment of the Iranians, and was occasionally mentioned in the works of other people. With the advent of the Ilkhanate, the idea experienced a resurgence.

Mustawfi describes the borders of Iran extending from the Indus River to Khwarazm and Transoxiana in the east to Byzantium and Syria in the west, corresponding to the territory of the Sasanian Empire. He defines the provinces of Iran in 20 chapters:

> Arab Iraq, Persian Iraq, Arran, Mughan, Shirvan, Georgia, Byzantium, Armenia, Rabi'a, Kurdistan, Khuzestan, Fars, Shabankara, Kirman, Mukran, Hormuz, Nimruz, Khorasan, Mazandaran, Qumis, Tabaristan and Gilan.

This way of conceptualizing the history and geography of Iran has been emulated by other historians since the 13th-century.

Biographical Summary

A native of Qazvin, Mustawfi belonged to family of mustawfis (financial accountants), thus his name. He was a close associate of the prominent vizier and historian Rashid al-Din Hamadani, who inspired him to write historical and geographical works. Mustawfi is the author of three works: (1) Tarikh-i guzida ("Excerpt History"), (2) Zafarnamah ("Book of Victory") and (3) Nuzhat al-qulub ("Hearts' Bliss"). A highly influential figure, Mustawfi's way of conceptualizing the history and geography of Iran has been emulated by other historians after him.

He is buried in a dome-shaped mausoleum in his native Qazvin.

Mustawfi was born in 1281 in the town of Qazvin, located in Persian Iraq (Irāq-i Ajam), a region corresponding to the western part of Iran. His family was descended from Arabs that had occupied the governorship of the town in the 9th and 10th-centuries, later to serve as mustawfis (high-ranking financial accountants) at the advent of the Ghaznavids. Mustawfi's great-grandfather Amin al-Din Nasr had served as the mustawfi of Iraq, which had since then been the moniker of the family. Amin al-Din Nasr, during his retirement, was killed in 1220 by raiding Mongols after the sack of Qazvin, during the Mongol invasion of Iran.

Regardless, Mustawfi's family still greatly served the Mongols and even rose to further prominence during this period; his older cousin Fakhr al-Din Mustawfi briefly served as vizier of the Ilkhanate, while his brother Zayn al-Din was an assistant of the prominent vizier and historian, Rashid al-Din Hamadani. Mustawfi's family was thus amongst those many families from Persian Iraq who rose to prominence during the Mongol era. Rivalry soon arose between the Persian Iraqis and the already established Khurasanis, particularly between the Mustawfis and the Juvayni families. Mustawfi was appointed, in 1311, as the financial accountant of his native Qazvin, as well as other neighboring districts, including Abhar, Zanjan and Tarumayn.

He had been appointed to this post by Rashid al-Din, who made him gain an interest in history, inspiring him to start writing the Zafarnamah ("Book of Victory") in 1320, as a continuation of Ferdowsi's Shahnameh ("Book of Kings"). He completed the work in 1334, consisting of 75,000 verses, reporting the history of the Islamic era up until the Ilkhanate era.

Before that, he had also written a similar chronicle titled Tarikh-i guzida ("Excerpt History") in 1330, which was his first work. The chronicle, made for Rashid al-Din's son Ghiyath al-Din Muhammad, was a world history, narrating the events of the prophets, the pre-Islamic kings of Iran, and the Islamic world.

Nothing is known of Mustawfi's life during the end of the Ilkhanate, except that he travelled between Tabriz and Baghdad.

In the summer of 1339, Mustawfi was at Sawa, working for Ghiyath al-Din Muhammad's son-in-law Hajji Shams al-Din Zakariya, who was the vizier of the Jalayirid ruler, Hasan Buzurg (r. 1336–1356). There he tried to help with the management of the divan, but soon found himself unemployed after Hasan Buzurg's retreat to Baghdad due to a defeat by the Chobanid prince, Hasan Kuchak.

Mustawfi was ambivalent whether to return to his native Qazvin or flee to the much more secure southern Iran. He eventually chose to leave for the southern Iranian city of Shiraz to seek better fortunes, but was let down by the reception he received at the court of the Injuid ruler Amir Mas'ud Shah (r. 1338–1342).

Nevertheless, he remained there for ten months more, until he chose to leave due to the chaos that ensued during the Injuid dynastic struggle for the throne. He returned up north, where he was well received in Awa, Sawa, Kashan and Isfahan, finally returning to Qazvin at the end of 1340. He mentions the turmoil he went through during this period in several of his poems, and also went through illness, until he recouped after gaining sympathy from an unknown patron, possibly

Hasan Buzurg. It was around this time that Mustawfi completed his cosmographical and geographical work Nuzhat al-Qulub ("Hearts' Bliss").

He died sometime after 1339/40 in Qazvin, where he was buried in a dome-shaped mausoleum.

66. Shihab al-Umari

Shihab al-Din Abu al-Abbas Ahmad ibn Fadlallah al-Umari (Arabic: شهاب الدين أبو العبّاس أحمد بن فضل الله العمري), commonly known as Ibn Fadlallah al-Umari or al-Umari, (1300 – 1349), was a historian born in Damascus.

Contributions in Social Sciences

His major works include (1) at-Taʿrīf bi-al-muṣṭalaḥ ash-sharīf, on the subject of the Mamluk administration, and (2) Masālik al-abṣār fī mamālik al-amṣār, an encyclopedic collection of related information. The latter was translated into French by Maurice Gaudefroy-Demombynes in 1927.

Ibn Fadlallah visited Cairo shortly after the Malian Mansa Kankan Musa I's pilgrimage to Mecca. Therefore, his writings are a primary source for this legendary hajj of the Malian Mensa. This Haj is of special importance because it connects with the falsification of the myth that Columbus discovered America.

Ibn Fadlallah provides evidence for such a falsification. First, he recorded that Kankan Musa asserted the following about the previous Mansa:

Kankan Musa stated that the previous ruler had abdicated the throne to journey to a land across the ocean.

A Malian King was much richer than Portugal was in 15th century; and could certainly have taken up such a project with ease. Ibn Fadlallah records about the wealth of a Malian King:

> Mansa Kankan Musa dispensed so much gold that its value fell in Egypt for a decade afterward.

This evidence led contemporary Malian historian Gaoussou Diawara to assert that a Malian Mansa (perhaps Abu Bakr II) reached the Americas over a century and a half before Christopher Columbus did.

The works of Ibn Fadlallah also provide a basis for the Muslim side on the wars of Amda Seyon I, that took place against Ifat, Adal, and other regions.

Biographical Summary

Ibn Fadlallah was a student of Ibn Taymiyya. He was born in 1300 AD and he died in 1349.

67. Ibn Juzayy

Abu al-Qasim, Muhammad b. Ahmad b. Muhammad b. 'Abd Allah, Ibn Juzayy al-Kalbi al-Gharnati

(Arabic: أبو القاسم، محمد بن أحمد بن محمد بن عبد الله، ابن جزي الكلبي الغرناطي), was an Andalusian Maliki-Ash'ari scholar and poet.

Contributions in Social Sciences

Ibn Juzayy is mainly known as the writer to whom Ibn Battuta dictated an account of his travels. He wrote "The Travels of Ibn Battuta" (Riḥlat Ibn Baṭūṭah) in 1352-55.

Of course, he wrote other treatises:

- Ibn Juzayy wrote the treatise titled “al-Qawanin al-Fiqhiyyah” or "The Laws of Jurisprudence". It is a comparative manual of the jurisprudence of the four Sunni madhhabs (Maliki, Hanafi, Shafi`i, Hanbali) with emphasis on the Maliki school and notices of the views of the Ẓāhirī school and others.
- He is also noted for his tafsir of the Qur'an titled “al-Tashil li Ulum al-Tanzil”.
- He wrote his book on legal theory “Taqrīb al-Wuṣūl ‘ilā ‘Ilm al-Uṣūl” or The Nearest of Paths to the Knowledge of the Fundamentals of Islamic Jurisprudence. He wrote this book for his son.
- He wrote a treatise on Sufism based on the Qur'an, “The Refinement of the Hearts”.

Biographical Summary

Ibn Juzayy was an Andalusian. He was a Maliki-Ash'ari scholar and a poet. Ibn Juzayy died in 1357.

68. Ibn Battuta

Shams al-Din Abu'Abdallah Muhammad ibn'Abdallah ibn Muhammad ibn Ibrahim ibn Muhammad ibn Yusuf Lawati al-Tanji ibn Battuta.

24 February 1304 – 1368/1369),

commonly known as Ibn Battuta,

was a Muslim traveler, explorer and scholar. Over a period of thirty years from 1325 to 1354, Ibn Battuta visited most of North Africa, the Middle East, East Africa, Central Asia, South Asia, Southeast Asia, China, the Iberian Peninsula, and West Africa. Ibn Battuta travelled more than any other explorer in pre-modern history, totaling around 117,000 km (73,000 mi), surpassing Zheng He with about 50,000 km (31,000 mi) and Marco Polo with 24,000 km (15,000 mi).

Near the end of his life, he dictated an account of his journeys, titled A Gift to Those Who Contemplate the Wonders of Cities and the Marvels of Travelling, but commonly known as The Rihla.

Contributions in Social Sciences

Ibn Battuta is known worldwide for his journeys; these are most extensive in the world history until before the modern times of trains, cars and planes. His travel accounts were dictated to Ibn Juzayy who is discussed in the previous entry in this book.

Travels

First Pilgrimage

On 2 Rajab in the Muslim year 725 Hijrah (14 June 1325 AC), at the age of twenty-one, Ibn Battuta set off from his home town on a hajj, a journey that would ordinarily have taken sixteen months. He was eager to learn more about far-away lands and craved adventure. He would not return to Morocco again for 24 years.

> I set out alone, having neither fellow-travelers in whose companionship I might find cheer, nor caravan whose part I might join, but swayed by an overmastering impulse within me and a desire long-cherished in my bosom to visit these illustrious sanctuaries. So, I braced my resolution to quit my dear ones, female and male, and forsook my home as birds forsake their nests. My parents being yet in the bonds of life, it weighed sorely upon me to part from them, and both they and I were afflicted with sorrow at this separation.

He travelled to Mecca overland, following the North African coast across the sultanates of Abd al-Wadid and Hafsid. The route took him through Tlemcen, Béjaïa, and then Tunis, where he stayed for two months. For safety, Ibn Battuta usually joined a caravan to reduce the risk of being robbed. He took a bride in the town of Sfax, but soon left her due to a dispute with the father. That was the first in a series of marriages that would feature in his travels.

In the early spring of 1326, after a journey of over 3,500 km (2,200 mi), Ibn Battuta arrived at the port of Alexandria, at the time part of the Bahri Mamluk empire. He met two ascetic pious men in Alexandria.

One was Sheikh Burhanuddin, who is supposed to have foretold the destiny of Ibn Battuta as a world traveler and told him, "It seems to me that you are fond of foreign travel. You must visit my brother Fariduddin in India, Rukonuddin in Sind, and Burhanuddin in China. Convey my greetings to them."

Another pious man, Sheikh Murshidi, interpreted the meaning of a dream of Ibn Battuta as being that he was meant to be a world traveler.

He spent several weeks visiting sites in the area, and then headed inland to Cairo, the capital of the Mamluk Sultanate. After spending about a month in Cairo, he embarked on the first of many detours within the relative safety of Mamluk territory. Of the three usual routes to Mecca, Ibn Battuta chose the least-travelled, which involved a journey up the Nile valley, then east to the Red Sea port of ʿAydhab. Upon approaching the town, however, a local rebellion forced him to turn back.

Ibn Battuta returned to Cairo and took a second side trip, this time to Mamluk-controlled Damascus. During his first trip he had encountered a holy man who prophesied that he would only reach Mecca by travelling through Syria. The diversion held an added advantage; because of the holy places that lay along the way, including Hebron, Jerusalem, and Bethlehem, the Mamluk authorities kept the route safe for pilgrims. Without this help many travelers would be robbed and murdered.

After spending the month of Ramadan, during August, in Damascus, he joined a caravan travelling the 1,300 km (810 mi) south to Medina, site of the Mosque of the Islamic prophet Muhammad. After four days in the town, he journeyed on to Mecca while visiting holy sites along the way; upon his arrival to Mecca he completed his first pilgrimage in November, and he took the honorific status of El-Hajji. Rather than returning home, Ibn Battuta decided to continue travelling, choosing as his next destination the Ilkhanate, a Mongol Khanate, to the northeast.

Iraq and Iran

On 17 November 1326, following a month spent in Mecca, Ibn Battuta joined a large caravan of pilgrims returning to Iraq across the Arabian Peninsula. The group headed north to Medina and then, travelling at night, turned northeast across the Najd plateau to Najaf, on a journey that lasted about two weeks. In Najaf, he visited the mausoleum of Ali, the Fourth Caliph.

Then, instead of continuing to Baghdad with the caravan, Ibn Battuta started a six-month detour that took him into Iran. From Najaf, he journeyed to Wasit, then followed the river Tigris south to Basra. His next destination was the town of Isfahan across the Zagros Mountains in Iran. He then headed south to Shiraz; a large, flourishing city spared the destruction wrought by Mongol invaders on many more northerly towns. Finally, he returned across the mountains to Baghdad, arriving there in June 1327. Parts of the city were still ruined from the damage inflicted by Hulagu Khan's invading army in 1258.

In Baghdad, he found Abu Sa'id, the last Mongol ruler of the unified Ilkhanate, leaving the city and heading north with a large retinue. Ibn Battuta joined the royal caravan for a while, then turned north on the Silk Road to Tabriz, the first major city in the region to open its gates to the Mongols and by then an important trading center as most of its nearby rivals had been razed by the Mongol invaders.

Ibn Battuta left again for Baghdad, probably in July, but first took an excursion northward along the river Tigris. He visited Mosul, where he was the guest of the Ilkhanate governor. and then the towns of Cizre (Jazirat ibn 'Umar) and Mardin in modern-day Turkey. At a hermitage on a mountain near Sinjar, he met a Kurdish mystic who gave him some silver coins. Once back in Mosul, he joined a "feeder" caravan of pilgrims heading south to Baghdad, where they would meet up with the main caravan that crossed the Arabian Desert to Mecca. Ill with diarrhea, he arrived in the city weak and exhausted for his second hajj.

Arabia

Ibn Battuta remained in Mecca for some time (the Rihla suggests about three years, from September 1327 until autumn 1330). Problems with chronology, however, lead commentators to suggest that he may have left after the 1328 hajj.

After the hajj in either 1328 or 1330, he made his way to the port of Jeddah on the Red Sea coast. From there he followed the coast in a series of boats (known as a jalbah, these were small craft made of wooden planks sewn together, lacking an established phrase) making slow progress against the prevailing south-easterly winds. Once in Yemen he visited Zabīd and later the highland town of Ta'izz, where he met the Rasulid dynasty king (Malik) Mujahid Nur al-Din Ali. Ibn Battuta also mentions visiting Sana'a. He went to the important trading port of Aden, arriving around the beginning of 1331.

Somalia

From Aden, Ibn Battuta embarked on a ship heading for Zeila on the coast of Somalia. He then moved on to Cape Guardafui further down the Somali seaboard, spending about a week in each location. Later he would visit Mogadishu, the then pre-eminent city of the "Land of the Berbers" (بلد البربر Balad al-Barbar, the medieval Arabic term for the Horn of Africa).

When Ibn Battuta arrived in 1332, Mogadishu stood at the zenith of its prosperity. He described it as "an exceedingly large city" with many rich merchants, noted for its high-quality fabric that was exported to other countries, including Egypt. Ibn Battuta added that the city was ruled by a Somali Sultan, Abu Bakr ibn Shaikh 'Umar. He noted that Sultan Abu Bakr had dark skin complexion and spoke in his native tongue (Somali), but was also fluent in Arabic. The Sultan also had a retinue of wazirs (ministers), legal experts, commanders, royal eunuchs, and other officials at his beck and call.

Swahili Coast

Ibn Battuta continued by ship south to the Swahili coast, a region then known in Arabic as the Bilad al-Zanj ("Land of the Zanj") with an overnight stop at the island town of Mombasa. Although relatively small at the time, Mombasa would become important in the following century. After a journey along the coast, Ibn Battuta next arrived in the island town of Kilwa in present-day Tanzania, which had become an important transit center of the gold trade. He described the city as "one of the finest and most beautifully built towns; all the buildings are of wood, and the houses are roofed with dīs reeds".

Ibn Battuta recorded his visit to the Kilwa Sultanate in 1330, and commented favorably on the humility and religion of its ruler, Sultan al-Hasan ibn Sulaiman, a descendant of the legendary Ali ibn al-Hassan Shirazi. He further wrote that the authority of the Sultan extended from Malindi in the north to Inhambane in the south and was particularly impressed by the planning of the city, believing it to be the reason for Kilwa's success along the coast. During this period, he described the construction of the Palace of Husuni Kubwa and a significant extension to the Great Mosque of Kilwa, which was made of coral stones and was the largest mosque of its kind. With a change in the monsoon winds, Ibn Battuta sailed back to Arabia, first to Oman and the Strait of Hormuz then on to Mecca for the hajj of 1332.

Anatolia

After his third pilgrimage to Mecca, Ibn Battuta decided to seek employment with the Sultan of Delhi, Muhammad bin Tughluq. In the autumn of 1332, he set off for the Seljuk controlled territory of Anatolia to take an overland route to India. He crossed the Red Sea and the Eastern Desert to reach the Nile valley and then headed north to Cairo. From there he crossed the Sinai Peninsula to Palestine and then travelled north again through some of the towns that he had visited in 1326. From the Syrian port of Latakia, a Genoese ship took him to Alanya on the southern coast of modern-day Turkey.

He then journeyed westwards along the coast to the port of Antalya. In the town he met members of one of the semi-religious fityan associations. These were a feature of most Anatolian towns in the 13th and 14th centuries. The members were young artisans and had at their head a leader with the title of Akhil. The associations specialized in welcoming travelers. Ibn Battuta was very impressed with the hospitality that he received and would later stay in their hospices in more than 25 towns in Anatolia. From Antalya Ibn Battuta headed inland to Eğirdir which was the capital of the Hamidids. He spent Ramadan (May 1333) in the city.

From this point his itinerary across Anatolia in the Rihla becomes confused. Ibn Battuta describes travelling westwards from Eğirdir to Milas and then skipping 420 km (260 mi) eastward past Eğirdir to Konya. He then continues travelling in an easterly direction, reaching Erzurum from where he skips 1,160 km (720 mi) back to Birgi which lies north of Milas. Historians believe that Ibn Battuta visited a number of towns in central Anatolia, but not in the order in which he describes. This could be for many reasons: Ibn Battutab may not have recalled the details, he may have regarded such details as not noteworthy, or the scribe may have omitted them for some reasons.

When Ibn Battuta arrived in Iznik, it had just been conquered by Orhan, Sultan of the nascent Ottoman Empire. Orhan was away and his wife was in command of the nearby stationed soldiers, Ibn Battuta gave this account of Orhan's wife:

> "A pious and excellent woman. She treated me honorably, gave me hospitality and sent gifts."
>
> Ibn Battuta's account of Orhan:
>
> The greatest of the kings of the Turkmens and the richest in wealth, lands and military forces. Of fortresses, he possesses nearly a hundred, and for most of his time, he is continually engaged in making a round of them, staying in each fortress for some days to put it in good order and examine its condition. It is said that he has never stayed for a whole month in any one town. He also fights with the infidels continually and keeps them under siege.
>
> —Ibn Battuta

Ibn Battuta had also visited Bursa which at the time was the capital of the Ottoman Beylik, he described Bursa as

> "a great and important city with fine bazaars and wide streets, surrounded on all sides with gardens and running springs".

Central Asia

From Sinope he took a sea route to the Crimean Peninsula, arriving in the Golden Horde realm. He went to the port town of Azov, where he met with the emir of the Khan, then to the large and rich city of Majar. He left Majar to meet with Uzbeg Khan's travelling court (Orda), which was at the time near Mount Beshtau. From there he made a journey to Bolghar, which became the northernmost point he reached, and noted its unusually short nights in summer (by the standards of the subtropics). Then he returned to the Khan's court and with it moved to Astrakhan.

Ibn Battuta recorded that while in Bolghar he wanted to travel further north into the land of darkness. The land is snow-covered throughout (northern Siberia) and the only means of transport is dog-drawn sled. There lived a mysterious people who were reluctant to show themselves. They traded with southern people in a peculiar way. Southern merchants brought various goods and placed them in an open area on the snow in the night, then returned to their tents. Next morning they came to the place again and found their merchandise taken by the mysterious people, but in exchange they found fur-skins which could be used for making valuable coats, jackets, and other winter garments. The trade was done between merchants and the mysterious people without seeing each other. As Ibn Battuta was not a merchant and saw no benefit of going there, he abandoned the travel to this land of darkness.

When they reached Astrakhan, Öz Beg Khan had just given permission for one of his pregnant wives, Princess Bayalun. She was a daughter of Byzantine emperor Andronikos III Palaiologos. She was given permission to return to her home city of Constantinople to give birth.

Ibn Battuta talked his way into this expedition. This was his first travel beyond the boundaries of the Islamic world.

Arriving in Constantinople towards the end of 1334, he met the Byzantine emperor Andronikos III Palaiologos. He visited the great church of Hagia Sophia and spoke with an Eastern Orthodox priest about his travels in the city of Jerusalem. After a month in the city, Ibn Battuta returned to Astrakhan, and then arrived in the capital city Sarai al-Jadid, where he reported the accounts of his travels to Sultan Öz Beg Khan (r. 1313–1341).

Then he continued past the Caspian and Aral Seas to Bukhara and Samarkand, the latter of which he praised as "one of the grandest and finest cities, and the most perfect of them". Here he visited the court of another Mongol khan, Tarmashirin (r. 1331–1334) of the Chagatai Khanate. He also noted the ruined state of the city walls, a result of the Mongol invasion in 1220 and subsequent infighting. From there, he journeyed south to Afghanistan, then crossed into India via the mountain passes of the Hindu Kush. In the Rihla, he mentions these mountains and the history of the range. He wrote,

> After this I proceeded to the city of Barwan, in the road to which is a high mountain, covered with snow and exceedingly cold; they call it the Hindu Kush, that is Hindu-killer, because most of those arriving thither from India die on account of the intenseness of the cold.

—Ibn Battuta, Chapter XIII, Rihla – Khorasan

Ibn Battuta and his party reached the Indus River on 12 September 1333. From there, he made his way to Delhi and became acquainted with the sultan, Muhammad bin Tughluq.

South Asia

Muhammad bin Tughluq patronized various scholars, Sufis, qadis, viziers, and other functionaries in order to consolidate his rule. On the strength of his years of study in Mecca, Ibn Battuta was appointed a qadi, or judge, by the sultan. However, he found it difficult to enforce Islamic law beyond the sultan's court in Delhi, due to lack of Islamic appeal in India.

It is uncertain by which route Ibn Battuta entered the Indian subcontinent but he was kidnapped and robbed by rebels on his journey to the Indian coast. He may have entered via the Khyber Pass and Peshawar, or further south. He crossed the Sutlej river near the city of Pakpattan, in modern-day Pakistan, where he paid obeisance at the shrine of Baba Farid, before crossing southwest into Rajput country. From the Rajput kingdom of Sarsatti, Ibn Battuta visited Hansi in India, describing it as "among the most beautiful cities, the best constructed and the most populated; it is surrounded with a strong wall, and its founder is king Tara". Upon his arrival in Sindh, Ibn Battuta mentions the Indian rhinoceros that lived on the banks of the Indus.

The Sultan was erratic even by the standards of the time and for six years Ibn Battuta veered between living the high life of a trusted subordinate and falling under suspicion of treason for a variety of offences. His plan to leave on the pretext of taking another hajj was stymied by the Sultan. The opportunity for Ibn Battuta to leave Delhi finally arose in 1341 when an embassy arrived from the Yuan dynasty of China asking for permission to rebuild a Himalayan Buddhist temple, popular with Chinese pilgrims.

Ibn Battuta was given charge of the embassy but en route to the coast at the start of the journey to China, he and his large retinue were attacked by a group of bandits. Separated from his companions, he was robbed, kidnapped, and nearly lost his life. Despite this setback, within ten days he had caught up with his group and continued on to Khambhat in the Indian state of Gujarat. From there, they sailed to Calicut (now known as Kozhikode), where Portuguese explorer Vasco da Gama would land two centuries later. While in Calicut, Ibn Battuta was the guest of the ruling Zamorin. While Ibn Battuta visited a mosque on shore, a storm arose and one of the ships of his expedition sank. The other ship then sailed without him only to be seized by a local Sumatran king a few months later.

Afraid to return to Delhi and be seen as a failure, he stayed for a time in southern India under the protection of Jamal-ud-Din, ruler of the small but powerful Nawabi sultanate on the banks of the Sharavathi river next to the Arabian Sea. This area is today known as Hosapattana and lies in the Honavar administrative district of Uttara Kannada. Following the overthrow of the sultanate, Ibn Battuta had no choice but to leave India. Although determined to continue his journey to

China, he first took a detour to visit the Maldive Islands where he had worked as a judge.

He spent nine months on the islands, much longer than he had intended. When he arrived at the capital, Malé, Ibn Battuta did not plan to stay. However, the leaders of the formerly Buddhist nation that had recently converted to Islam were looking for a chief judge, someone who knew Arabic and the Qur'an. To convince him to stay they gave him pearls, gold jewelry, and slaves, while at the same time making it impossible for him to leave by ship. Compelled into staying, he became a chief judge and married into the royal family of Omar I.

From the Maldives, he carried on to Sri Lanka and visited Sri Pada and Tenavaram temples. Ibn Battuta's ship almost sank on embarking Sri Lanka; the vessel that came to the rescue was attacked by pirates. Stranded onshore, he worked his way back to the Madurai kingdom in India. Here he spent some time in the court of the short-lived Madurai Sultanate under Ghiyas-ud-Din Muhammad Damghani. From here he returned to the Maldives and boarded a Chinese junk, still intending to reach China and take up his ambassadorial post.

He reached the port of Chittagong in modern-day Bangladesh intending to travel to Sylhet to meet Shah Jalal, who became so renowned that Ibn Battuta, then in Chittagong, made a one-month journey through the mountains of Kamaru near Sylhet to meet him. On his way to Sylhet, Ibn Battuta was greeted by several of Shah Jalal's disciples who had come to assist him on his journey many days before he had arrived. At the meeting in 1345 CE, Ibn Battuta noted that Shah Jalal was tall and lean, fair in complexion and lived by the mosque in a cave, where his only item of value was a goat he kept for milk, butter, and yogurt. He observed that the companions of the Shah Jalal were foreign and known for their strength and bravery. He also mentions that many people would visit the Shah to seek guidance. Ibn Battuta went further north into Assam, then turned around and continued with his original plan.

Southeast Asia

In 1345, Ibn Battuta traveled to Samudra Pasai Sultanate (called "al-Jawa") in present-day Aceh, Northern Sumatra, after 40 days voyage from Sunur Kawan. He notes in his travel log that the ruler of Samudra Pasai was a pious Muslim named Sultan Al-Malik Al-Zahir Jamal-ad-Din, who performed his religious duties with utmost zeal and often waged campaigns against animists in the region. The island of Sumatra, according to Ibn Battuta, was rich in camphor, areca nut, cloves, and tin.

The madh'hab he observed was that of Imam Al-Shafi'i, whose customs were similar to those he had previously seen in coastal India, especially among the Mappila Muslims, who were also followers of Imam Al-Shafi'i. At that time Samudra Pasai marked the end of Dar al-Islam, because no territory east of this was ruled by a Muslim. Here he stayed for about two weeks in the wooden walled town as a guest of the sultan, and then the sultan provided him with supplies and sent him on his way on one of his own junks to China.

Ibn Battuta first sailed for 21 days to a place called "Mul Jawa" (island of Java or Majapahit Java) which was a center of a Hindu empire. The empire spanned 2 months of travel, and ruled over the country of Qaqula and Qamara. He arrived at the walled city named Qaqula/Kakula, and observed that the city had war junks for pirate raiding and collecting tolls and that elephants were employed for various purposes. He met the ruler of Mul Jawa and stayed as a guest for three days.

Ibn Battuta then sailed to a state called Kaylukari in the land of Tawalisi, where he met Urduja, a local princess. Urduja was a brave warrior, and her people were opponents of the Yuan dynasty. She was described as an "idolater", but could write the phrase Bismillah in Islamic calligraphy.

The locations of Kaylukari and Tawalisi are not known in present day geography.

From Kaylukari, Ibn Battuta finally reached Quanzhou in Fujian Province, China.

China

In the year 1345, Ibn Battuta arrived at Quanzhou in China's Fujian province, then under the rule of the Mongol-led Yuan dynasty. One of the first things he noted was that Muslims referred to the city as "Zaitun" (meaning olive), but Ibn Battuta could not find any olives anywhere. He mentioned local artists and their mastery in making portraits of newly arrived foreigners; these were for security purposes. Ibn Battuta praised the craftsmen and their silk and porcelain; as well as fruits such as plums and watermelons and the advantages of paper money.

He described the manufacturing process of large ships in the city of Quanzhou. He also mentioned Chinese cuisine and its usage of animals such as frogs, pigs, and even dogs which were sold in the markets, and noted that the chickens in China were larger than those in the west.

In Quanzhou, Ibn Battuta was welcomed by the head of the local Muslim merchants (possibly a fānzhǎng or "Leader of Foreigners"; and Sheikh al-Islam (Imam)), who came to meet him with flags, drums, trumpets, and musicians. Ibn Battuta noted that the Muslim populace lived within a separate portion in the city where they had their own mosques, bazaars, and hospitals. In Quanzhou, he met two prominent Iranians, Burhan al-Din of Kazerun and Sharif al-Din from Tabriz (both of whom were influential figures noted in the Yuan History as "A-mi-li-ding" and "Sai-fu-ding", respectively). While in Quanzhou he ascended the "Mount of the Hermit" and briefly visited a well-known Taoist monk in a cave.

He then travelled south along the Chinese coast to Guangzhou, where he lodged for two weeks with one of the city's wealthy merchants.

From Guangzhou he went north to Quanzhou and then proceeded to the city of Fuzhou, where he took up residence with Zahir al-Din and met Kawam al-Din and a fellow countryman named Al-Bushri of Ceuta, who had become a wealthy merchant in China. Al-Bushri accompanied Ibn Battuta northwards to Hangzhou and paid for the gifts that Ibn Battuta would present to the Emperor Huizong of Yuan.

Ibn Battuta said that Hangzhou was one of the largest cities he had ever seen, and he noted its charm, describing that the city sat on a beautiful lake surrounded by gentle green hills. He mentions the city's Muslim quarter and resided as a guest with a family of Egyptian origin. During his stay at Hangzhou he was particularly impressed by the large number of well-crafted and well-painted Chinese wooden ships, with colored sails and silk awnings, assembling in the canals. Later he attended a banquet of the Yuan administrator of the city named Qurtai, who according to Ibn Battuta, was very fond of the skills of local Chinese conjurers. Ibn Battuta also mentions locals who worshipped a solar deity.

He described floating through the Grand Canal on a boat watching crop fields, orchids, merchants in black silk, and women in flowered silk and priests also in silk. In Beijing, Ibn Battuta referred to himself as the long-lost ambassador from Delhi Sultanate and was invited to the Yuan imperial court of Emperor Huizong (who according to Ibn Battuta was worshipped by some people in China). Ibn Batutta noted that the palace of Khanbaliq was made of wood and that the ruler's "head wife" (Empress Qi) held processions in her honor.

Ibn Battuta also wrote he had heard of "the rampart of Yajuj and Majuj" that was "sixty days' travel" from the city of Zeitun (Quanzhou). Ibn Battuta, who asked about the wall in China, could find no one who had either seen it or knew of anyone who had seen it.

Ibn Battuta travelled from Beijing to Hangzhou, and then proceeded to Fuzhou. Upon his return to Quanzhou, he soon boarded a Chinese junk owned by the Sultan of Samudera Pasai Sultanate heading for Southeast Asia, whereupon Ibn Battuta was unfairly charged a hefty sum by the crew and lost much of what he had collected during his stay in China.

Battuta claimed that the Emperor Huizong of Yuan had interred with him in his grave six slave soldiers and four girl slaves. Silver, gold, weapons, and carpets were put into the grave.

Return

After returning to Quanzhou in 1346, Ibn Battuta began his journey back to Morocco. In Kozhikode, he once again considered throwing himself at the mercy of Muhammad bin Tughluq in Delhi, but thought better of it and decided to carry on to Mecca. On his way to Basra he passed through the Strait of Hormuz, where he learned that Abu Sa'id, last ruler of the Ilkhanate Dynasty had died in Iran. Abu Sa'id's territories had subsequently collapsed due to a fierce civil war between the Iranians and Mongols.

In 1348, Ibn Battuta arrived in Damascus with the intention of retracing the route of his first hajj. He then learned that his father had died 15 years earlier, and death became the dominant theme for the next year or so. The Black Death had struck and he stopped in Homs as the plague spread through Syria, Palestine, and Arabia. He heard of terrible death tolls in Gaza, but returned to Damascus that July where the death toll had reached 2,400 victims each day. When he stopped in Gaza, he found it was depopulated. In Egypt he stayed at Abu Sir. Reportedly deaths in Cairo had reached levels of 1,100 each day. He made hajj to Mecca then he decided to return to Morocco, nearly a quarter of a century after leaving home. On the way he made one last detour to Sardinia, then in 1349, returned to Tangier by way of Fez, only to discover that his mother had also died a few months before.

Spain and North Africa

After a few days in Tangier, Ibn Battuta set out for a trip to the Muslim-controlled territory of al-Andalus on the Iberian Peninsula. King Alfonso XI of Castile and León had threatened to attack Gibraltar, so in 1350, Ibn Battuta joined a group of Muslims leaving Tangier with the intention of defending the port. By the time he arrived, the Black Death had killed Alfonso and the threat of invasion had receded, so he turned the trip into a travel tour ending up in Granada.

After his departure from al-Andalus he decided to travel through Morocco. On his return home, he stopped for a while in Marrakech, which was almost a ghost town following the recent plague and the transfer of the capital to Fez.

Mali and Timbuktu

In the autumn of 1351, Ibn Battuta left Fez and made his way to the town of Sijilmasa on the northern edge of the Sahara in present-day Morocco. There he bought a number of camels and stayed for four months. He set out again with a caravan in February 1352 and after 25 days arrived at the dry salt lake bed of Taghaza with its salt mines. All of the local buildings were made from slabs of salt by the slaves of the Masufa tribe, who cut the salt in thick slabs for transport by camel. Taghaza was a commercial center and awash with Malian gold, though

Ibn Battuta did not form a favorable impression of the place, recording that it was plagued by flies and the water was brackish.

After a ten-day stay in Taghaza, the caravan set out for the oasis of Tasarahla (probably Bir al-Ksaib) where it stopped for three days in preparation for the last and most difficult leg of the journey across the vast desert. From Tasarahla, a Masufa scout was sent ahead to the oasis town of Oualata, where he arranged for water to be transported a distance of four days travel where it would meet the thirsty caravan. Oualata was the southern terminus of the trans-Saharan trade route and had recently become part of the Mali Empire. Altogether, the caravan took two months to cross the 1,600 km (990 mi) of desert from Sijilmasa

From there, Ibn Battuta travelled southwest along a river he believed to be the Nile (it was actually the river Niger), until he reached the capital of the Mali Empire. There he met Mansa Suleyman, king since 1341. Ibn Battuta disapproved of the fact that female slaves, servants, and even the daughters of the sultan went about exposing parts of their bodies not befitting a Muslim. He wrote in his Rihla that black Africans were characterized by "ill manners" and "contempt for white men", and that he "was long astonished at their feeble intellect and their respect for mean things." He left the capital in February accompanied by a local Malian merchant and journeyed overland by camel to Timbuktu. Though in the next two centuries it would become the most important city in the region, at that time it was a small city and relatively unimportant. It was during this journey that Ibn Battuta first encountered a hippopotamus. The animals were feared by the local boatmen and hunted with lances to which strong cords were attached. After a short stay in Timbuktu, Ibn Battuta journeyed down the Niger to Gao in a canoe carved from a single tree. At the time Gao was an important commercial center.

After spending a month in Gao, Ibn Battuta set off with a large caravan for the oasis of Takedda. On his journey across the desert, he received a message from the Sultan of Morocco commanding him to return home. He set off for Sijilmasa in September 1353, accompanying a large caravan transporting 600 female slaves, and arrived back in Morocco early in 1354.

Ibn Battuta's itinerary gives scholars a glimpse as to when Islam first began to spread into the heart of west Africa.

Ibn Battuta's Travels Introduced to European Audience

After returning home from his travels in 1354, and at the suggestion of the Marinid ruler of Morocco, Abu Inan Faris, Ibn Battuta dictated an account in Arabic of his journeys to Ibn Juzayy, a scholar whom he had previously met in Granada. The account is the only source for Ibn Battuta's adventures. The full title of the manuscript is تحفة النظار في غرائب الأمصار وعجائب الأسفار, (Tuḥfat an-Nuẓẓār fī Gharāʾib al-Amṣār wa ʿAjāʾib al-Asfār), which can be translated as "A Masterpiece to Those Who Contemplate the Wonders of Cities and the Marvels of Travelling".) However, it is often simply referred to as The Travels (الرحلة, Rihla).

The introduction of Ibn Battuta's travel accounts to the European audience was gradual, over 19th and 20th centuries.

In the beginning of the 19th century, the German traveler-explorer Ulrich Jasper Seetzen (1767–1811) acquired a collection of manuscripts in the Middle East, among which was a 94-page volume containing an abridged version of Ibn Juzayy's text.

Three extracts were published in 1818 by the German writer Johann Kosegarten. A fourth extract was published the following year.

French writers were alerted to the initial publication by a lengthy review published in the Journal de Savants by Silvestre de Sacy.

Three copies of another abridged manuscript were acquired by the Swiss traveler, Johann Burckhardt, and bequeathed to the University of Cambridge. He gave a brief overview of their content in a book published posthumously in 1819.

The Arabic text was translated into English by Samuel Lee and published in London in 1829.

In the 1830s, during the French occupation of Algeria, the Bibliothèque Nationale (BNF) in Paris acquired five manuscripts of Ibn Battuta's travels, in which two were complete. One manuscript containing just the second part of the work is dated 1356; it is believed to be Ibn Juzayy's. The BNF manuscripts were used in 1843 by the Irish-French writer Baron de Slane to produce a translation into French of Ibn Battuta's visit to Sudan. They were also studied by the French writers Charles Defrémery and Beniamino Sanguinetti. Beginning in 1853 they published a series of four volumes containing a critical edition of the Arabic text together with a translation into

French. In their introduction Defrémery and Sanguinetti praised Lee's annotations but were critical of his translation which they claimed lacked precision, even in straightforward passages.

In 1929, exactly a century after the publication of Lee's translation, the historian Hamilton Gibb published an English translation of selected portions of Defrémery and Sanguinetti's Arabic text. Gibb had proposed to the Hakluyt Society in 1922 that he should prepare an annotated translation of the entire Rihla into English. His intention was to divide the translated text into four volumes, each volume corresponding to one of the volumes published by Defrémery and Sanguinetti. The first volume was not published until 1958. Gibb died in 1971, having completed the first three volumes. The fourth volume was prepared by Charles Beckingham and published in 1994.

Defrémery and Sanguinetti's printed text has now been translated into number of other languages.

Biographical Summary

All that is known about Ibn Battuta's life comes from the autobiographical information included in the account of his travels, which records that he was of Berber descent, born into a family of Islamic legal scholars (known as qadis) in Tangier on 24 February 1304, during the reign of the Marinid dynasty. His family belonged to a Berber tribe known as the Lawata. As a young man, he would have studied at a Maliki madhhab, the dominant form of education in North Africa at that time. Maliki Muslims requested that Ibn Battuta serve as their religious judge, as he was from an area where it was practiced.

He was born 24 February 1304 and died in 1368.

69. Ibn al-Khatib

Muhammad ibn Abdallah ibn Said ibn Ali ibn Ahmad al-Salmani

(Arabic: محمد بن عبدالله بن سعيد بن علي بن أحمدالسّلماني),

Also known as Ibn al-Khatib (Arabic: لسان الدين ابن الخطيب),

(Born 16 November 1313, Loja – died 1374, Fes),

was an Andalusi polymath: poet, writer, historian, philosopher, physician and politician from Emirate of Granada.

Contributions in Social Sciences

Ibn Al-Khatib is known for composing the muwashahs entitled "Jadaka al-Ghaithu" and "Lamma Bada Yatathanna."

He is highly esteemed both as a historian and as a poet. He was a contemporary of Ibn Khaldun.

His great historical work, al-Ihata fi akhbar Gharnata الإحاطة في أخبار غرناطة (The Complete Source on the History of Granada), written in 1369, includes his autobiography.

In his treatise about the plague (Muqni'at al-Sā'il 'an al-Maraḍ al-Hā'il), (ca.1362), Ibn al-Khatib presented the idea of transmission of disease through contagion. *That was 500 years before Louis Pasteur in France to whom the Europeans like to attribute it.* The original Arabic text is preserved in the Zaydani Collection at the Biblioteca del Real Monasterio de El Escorial, MS Arabic 1785.

In his treatise On the Plague, the physician Ibn al-Khatib writes:

"The existence of contagion is established by experience [and] by trustworthy reports on transmission by garments, vessels, ear-rings;

by the spread of it by persons from one house, by infection of a healthy sea-port by an arrival from an infected land [and] by the immunity of isolated individuals."

This research precedes centuries before Louis Pasteur conducted his research in Europe.

His poetry was influenced by court poets from the Mashreq, or Islamic east, especially Abū Nuwās, Abū Tammām, and al-Mutanabbī. Ibn al-Khatib was a master of saj' (سجع), or rhymed prose, especially in his maqamaat.

Some of his poems decorate the walls of the palace of Alhambra in Granada.

Following is a partial list of works by Ibn Al-Khatib:

- Danger During a Trip of Winter and Summer (خطرة الطيف في رحلة الشتاء والصيف): a description of a 21-day journey from Granada to Almería with Yusuf I, Sultan of Granada, composed in rhyming couplets.
- The Measurement of Choice in the Conditions of Places and Buildings (معيار الاختيار في ذكر المعاهد والديار): a muqama in which Ibn al-Khatib describes 34 Andalusi cities, including Malaga, Granada, and Ronda, comparing them to the Moroccan cities Tangier, Meknes, Fes, and Sebta, which he visited while exiled in Morocco.
- The Shaking of the Bag for Entertainment While Abroad (نفاضة الجراب في علالة الاغتراب): a collection of praise poetry, writings on history and geography, and personal narrative on his journey from the High Atlas back to Al-Andalus.

- The Badr View in the Nasirian State - Al-Lamhat al-Badriya fi al-Dawla al-Nasriya, ed. Arab & Latin transl. M.Casiri, Biblioteca arabico-hispana escurialensis, II, Madrid 1770.
- Compendium on Granada (in 5 vols.) - Al'Ihatat fi 'Akhbar Ghurnata (Arabic).
- Muqni'at al-Sā'il 'an al-Maraḍ al-Hā'il (مقنعة السائل عن المرض الهائل), a treatise on the Black Death and contagion, Zaydani Collection at the Biblioteca del Real Monasterio de El Escorial, MS Arabic 1785.
- The Scholars' Recitations of Dreams of the Kings of Islam.
- Biographies, Dates and Connections - 'Awsaf an-Naas fi al-Tawarikh wa'l-Salaat (Arabic).
- A Clerk after the People Move (Politics of Granada & Morocco) - Kanasat al-Dukan baad Intaqal as-Sakan.
- Calibrate Selection in Institutes of Mind
- Views of Sanseddin Ben Khatib in Morocco and Andalusia.
- Malaga and Sala
- The Masterpiece of the Book and the Purity of the Elect Manuscripts.
- Magic and Poetry.
- The Book of Rehana and the sorrow of the Creator.
- Garden Definition of Sharif Love.
- A Message in Politics.

Following is a reference to modern articles on the works by Ibn Al-Khatib:

- Jaysh Al-Tawshih of Lisan Al-Din Ibn Al-Khatib (Arabic), An Anthology of Andalusian Arabic Muwashshahat, Alan Jones (Editor), 1997 - ISBN 978-0-906094-42-6.
- Lisan Al Din Ibn Al Khatib, Tarikh Isbaniya Al Islamiya (history of Muslim Spain), ed. by Levi-Provençal, new edition, Cairo, 2004.
- Lisan Al Din Ibn Al Khatib, Awsaf Al Nas (description of peoples), Cairo, 2002.
- Lisan Al Din Ibn Al Khatib, Khaṭrat al-ṭayf : riḥlāt fī al-Maghrib wa-al-Andalus, 1347–1362, 2003.
- Lisan Al Din Ibn Al Khatib, Nafadhat al-jirab (the Ashtray of the Socks).
- Lisan al-Din ibn al-Khatib homme de lettres et historien, by Abdelbaqui Benjamaa, (French) thesis, Universite de la Sorbonne Nouvelle Paris III, 1992 (microform).

Biographical Summary

Ibn al-Khatib was born in 1313 AD at Loja, near Granada. Shortly after his birth, his father was appointed to a high post at the court of Emir Ismail I in Granada.

After his father and older brother were killed in the Battle of Río Salado in 1340, Ibn al-Khatib was hired to work as a secretary for his former teacher Ibn al-Jayyab, vizier to Emir Yusuf I. Following Ibn al-Jayyab's early death from the plague, Ibn al-Khatib became vizier and head of the emiri chancery, serving also in diplomatic roles in the courts of Andalusi and Maghrebi rulers.

For much of his life he was vizier at the court of the Sultan of Granada, Muhammed V. He spent two periods in exile in the Marinid empire: between 1360 and 1362, and 1371–74, he resided variously at Ceuta, Tlemcen and Fes.

In 1374, he was imprisoned for 'Zandaqa (heresy). He was sentenced to death by suffocation. His body was burned before being buried at Bab Mahruq, a city gate in Fes.

A detailed account of his demise was written down by Ibn Khaldun.

As a loyal courtier of Muhammed V of Granada, Ibn al-Khatib was arrested in the wake of a 1359 coup by Muhammed's half-brother Ismail, and had his property confiscated. He was soon released due to interference by the Marinid sultan of Morocco and joined a host of Andalusian refugees in Morocco, where he settled in the Atlantic town of Salé.

Here, he immersed himself in Sufi mysticism and writing. It was during this stay in Morocco that he first met Ibn Khaldun, as well as other important North African intellectuals, such as Ibn Marzuq.

In 1362, the former emir of Granada, Muhammed V, was able to regain the throne with help from the Moroccan sultan. This allowed Ibn al-Khatib to return to Granada and resume the office of Great Vizier (dhu al-wizaratayn, i.e. 'possessor of the two vizierates', meaning 'head of both the civil and military authority').

He soon ran afoul of severe political intrigues. He was eventually able to strengthen his own position while organizing the expulsion from Granada of a number of his North African political rivals. This caused friction, within Granada, between supporters of the expelled North Africans at Granada and Ibn al-Khatib.

His intrigues had made him an unpopular figure in some circles, causing two of his own students, Ibn Zamrak and Ibn Farkun, to join hands with his most powerful enemy in Granada, the Grand Qadi al-Nubahi, a man who had long held a grudge against Ibn al-Khatib. More importantly, emir Muhammed V had grown distrustful of Ibn al-Khatib for his overbearing control of the Granadan state and his strict loyalty to the Marinids of Morocco.

Feeling the heat rise, in 1371 Ibn al-Khatib left for North Africa, where he was well received by the Marinid sultan Abu Faris Abdul Aziz I of Morocco.

During his refuge in Morocco, the Grand Qadi al-Nubahi of Granada issued a fatwa in which Ibn al-Khatib's work on Sufism and philosophy were branded heretical and his work ordered to be burned. Ibn al-Khatib wrote a refutation of the fatwa, in which he harshly attacked al-Nubahi.

Numerous attempts by Granada to get Ibn al-Khatib either extradited or executed were fruitless, as the Moroccan sultan Abu Faris Abdul Aziz I refused to do so.

Though the Moroccan sultan soon died, Ibn al-Khatib was ensured of protection from Ibn Ghazi, Morocco's main vizier.

Ibn al-Khatib's future turned bleak when a Granada-supported coup removed Ibn Ghazi from office and brought a new sultan to power, Abu'l-Abbas Ahmad al-Mustansir. Indebted to Granada, the new sultan ordered Ibn al-Khatib arrested and a trial be held in the

Moroccan capital city of Fes, in which a Granadan group of emissaries, including his former student Ibn Zamrak, was actively involved.

Despite intimidation and torture, Ibn al-Khatib kept protesting his innocence throughout the trial and denied the accusations of heresy. The final vote was far from unanimous and a council of Islamic scholars were unable to reach a conclusive decision.

He was sent back to his prison cell and strangled later that night. On the next morning his body was buried near Fes' Bab al-Mahruq city gate. Unsatisfied, his enemies ordered his body dug up and thrown in a bonfire.

70. Ibn Khaldun

Abū Zayd ‘Abd ar-Raḥmān ibn Muḥammad ibn Khaldūn al-Ḥaḍramī (Arabic: أبو زيد عبد الرحمن بن محمد بن خلدون الحضرمي),

(27 May 1332 – 17 March 1406),

was a polymath who is regarded as the founder of the proto-disciplines that would become historiography, sociology, economics, and demography.

Contributions in Social Sciences

Ibn Khaldūn is founder of the proto-disciplines that would become historiography, sociology, economics, and demography.

His best-known book is the Muqaddimah or Prolegomena ("Introduction"), which he wrote in six months as he states in his autobiography.

Kitāb al-‘Ibar, (full title: Kitāb al-‘Ibar wa-Dīwān al-Mubtada’ wa-l-Khabar fī Ta’rīkh al-‘Arab wa-l-Barbar wa-Man ‘Āṣarahum min Dhawī ash-Sha’n al-Akbār "Book of Lessons, Record of Beginnings and Events in the History of the Arabs and the Berbers and Their Powerful Contemporaries"): it began as a history of the Berbers and expanded to a universal history in seven books.

- Book 1; Al-Muqaddimah ('The Introduction'), a socio-economic-geographical universal history of empires, and the best known of his works.
- Books 2-5; World History up to the author's own time.

- Books 6-7; Historiography of the Berbers and the Maghreb. Ibn Khaldun departs from the classical style of Arab historians by synthesizing multiple (sometimes contradictory) sources, without citations. Al-'Ibar remains an invaluable source of Berber history.

Ibn Khaldun diverged from norms that Muslim historians followed and rejected their focus on the credibility of the transmitter and focused instead on the validity of the stories, and encouraged critical thinking.

However, without Isnad (chain of reports) there is no way to verify and validate the historical narratives of a historian. The use of Isnad is a wonderful invention by the Muslim historians to ascertain the validity of history. Other historians that do not use any such mechanism for verification and validation, leave the door wide open for distortions, prejudice, and even outright lies.

In abandoning the Isnad, Ibn Khaldun gave up the superior accuracy in historiography that they afford and adopted a style that can promote arbitrariness. This is what the Europeans adopted by regarding Ibn Khaldun as their founder of the proto-discipline that would become historiography. No wonder that histories written by the Europeans suffer from inadvertently or even willfully entering the doors that are thus left wide open for distortions, prejudice, and even outright lies, because they do not use isnad as mechanism for verification and validation.

Because of such reasons Ibn Khaldun is not regarded as the founder of historiography by the Muslims; on the contrary, he faced criticism from his contemporaries, particularly Ibn Hajar al-`Asqalani. These criticisms included accusations of inadequate historical knowledge, an inaccurate title, disorganization, and a style resembling that of the prolific Arab literature writer, Al-Jahiz.

According to the scholar Sati' al-Husri, the Muqaddimah may be read as a sociological work. The work is based around Ibn Khaldun's central concept of 'aṣabiyyah. This social cohesion arises spontaneously in tribes and other small kinship groups; and it can be intensified and enlarged by a religious ideology. Ibn Khaldun's analysis looks at how this cohesion carries groups to power. Yet the same contains within itself the seeds of the group's downfall, to be replaced by a new group dynasty or empire bound by a stronger cohesion.

The Muqaddimah can also be analyzed in terms of an economic theory. *Ibn Khaldun describes the economy as being composed of value-adding processes.* Thereby, labor and skill are added to techniques and crafts, as components of an economic theory. He calls for the creation of a science to explain society, and goes on to outline these ideas in the Al-Muqaddimah.

Ibn Khaldun states:

"*Civilization and its well-being, as well as business prosperity, depend on productivity and people's efforts in all directions in their own interest and profit*".

Ibn Khaldun outlines early theories of division of labor, taxes, scarcity, and economic growth. He was the first to study the origin and causes of poverty. He argued that:

Poverty was a result of the destruction of morality and human values. He also looked at what factors contribute to wealth such as consumption, government, and investment. This is arguably a formula for the present-day concept of a GDP (Gross Domestic Product) but in a more generalized and less restrictive way.

Ibn Khaldun also argued that:

Poverty was not necessarily a result of poor financial decision-making but of external consequences; and therefore, the government should be involved in alleviating poverty.

The individual being cannot by himself obtain all the necessities of life. All human beings must co-operate to that end in their civilization. But what is obtained by the cooperation of a group of human beings satisfies the need of a number many times greater than themselves.

Ibn Khaldun believed that the currency of a monetary system should have intrinsic value and therefore be made of gold and silver (such as the dirham). He emphasized that the weight and purity of these coins should be strictly followed: the weight of one dinar should be one mithqal (the weight of 72 grains of barley, roughly 4.25 grams); and the weight of 7 dinar (i.e. 7/10 of a dinar) should be equal to 7/10 of a mithqal or 2.96 grams.

In Sociology, Ibn Khaldun's epistemology divided science into two different categories, the religious science that regards the sciences of the Qur'an, and the non-religious science. He further classified the non-religious sciences into intellectual sciences such as logic, arithmetic, geometry, astronomy, etc. and auxiliary sciences such as language, literature, poetry, etc.

To Ibn Khaldun, the state was a necessity of human society to restrain injustice within the society, but the state means force, thus itself an injustice. All societies must have a state governing them in order to establish a society. He attempted to standardize the history of societies by identifying ubiquitous phenomena present in all societies. To him, civilization was a phenomenon that will be present as long as humans exist. He characterized the fulfillment of basic needs as the beginning of civilization. At the beginning, people will look for different ways of increasing productivity of basic needs and expansion will occur. Later the society starts becoming more sedentary and focuses more on crafting, arts and the more refined characteristics. By the end of a society, it will weaken, allowing another small group of individuals to come into control. The conquering group is described as an unsatisfied group within the society itself or a group of bandits that constantly attack other weaker or weakened societies.

In the Muqaddimah, Ibn Khaldun describes the beginnings, development, cultural trends and the fall of societies; thus, leading to the rise of a new society which would then follow the same trends in a continuous cycle.

Ibn Khaldun believed that too much bureaucracy, such as taxes and legislations, would lead to the decline of a society, because it would constrain the development of more specialized labor such as increase in scholars and development of different services.

Ibn Khaldun believed that bureaucrats cannot understand the world of commerce and do not possess the same motivation as a businessman.

Ibn Khaldun emphasizes human beings' faculty to think (fikr) as what determines human behavior and ubiquitous patterns. This faculty is also what inspires human beings to form into a social structure to co-operate in division of labor and organization. The fikr faculty is the supporting pillar for all philosophical aspects of Ibn Khaldun's theory related to human beings' spiritual, intellectual, physical, social and political tendencies.

Ibn Khaldun emphasized in his epistemology the important aspect that educational tradition plays to ensure the new generations of a civilization continuously improve in the sciences and develop culture. Ibn Khaldun argued that without the strong establishment of an educational tradition, it would be very difficult for the new generations to maintain the achievements of the earlier generations, let alone improve them.

Advancements in literary works such as poems and prose were another way to distinguish the achievement of a civilization. However, Ibn Khaldun believed that whenever the literary facet of a society reaches its highest levels it ceases to indicate societal achievements anymore, but is an embellishment of life. He suggested that an increase of scholars and the quality of knowledge is an indicator for achievement of a society in sciences. The literary productions would be indicated by the manifestation of prose, poems and the artistic enrichment of a society.

From other sources we know of several other works, primarily composed during the time he spent in North Africa and Al-Andalus. His first book, Lubābu l-Muhassal, a commentary on the Islamic theology of Fakhr al-Din al-Razi, was written at the age of 19 under the supervision of his teacher al-Ābilī in Tunis. A work on Sufism, Shifā'u l-Sā'il, was composed around 1373 in Fes, Morocco. Whilst at the court of Muhammed V, Sultan of Granada, Ibn Khaldūn composed a work on logic, ‘allaqa li-s-Sulṭān.

Following are some works of Ibn Khaldun:

- Kitāb al-‘Ibar wa-Dīwān al-Mubtada’ wa-l-Khabar fī Ta’rīkh al-‘Arab wa-l-Barbar wa-Man ‘Āṣarahum min Dhawī ash-Sha’n al-Akbār
- Lubābu-l-Muhassal fee Uswoolu-d-Deen
- Shifā'u-s-Sā'il
- ‘Al-Laqaw li-s-Sulṭān
- Ibn Khaldun. 1951 التعريف بإبن خلدون ورحلته غربا وشرقا Al-Ta‘rīf bi Ibn-Khaldūn wa Riħlatuhu Għarbān wa Sharqān. Published by Muħammad ibn-Tāwīt at-Tanjī. Cairo (Autobiography in Arabic).
- Ibn Khaldūn. 1958 The Muqaddimah : An introduction to history. Translated from the Arabic by Franz Rosenthal. 3 vols. New York: Princeton.
- Ibn Khaldūn. 1967 The Muqaddimah : An introduction to history. Trans. Franz Rosenthal, ed. N.J. Dawood. (Abridged).
- Ibn Khaldun, 1332–1406. 1905 'A Selection from the Prolegomena of Ibn Khaldūn'. Trans. Duncan Macdonald

Biographical Summary

He is generally known as "Ibn Khaldūn" after a remote ancestor. His life is relatively well-documented, as he wrote an autobiography

"Presenting Ibn Khaldun and his Journey West and East" (التعريف بابن خلدون ورحلته غربا وشرقا, at-Taʿrīf bi-ibn Khaldūn wa-Riḥlatih Gharban wa-Sharqan). In this, numerous documents regarding his life are quoted word-for-word.

His full name is Abdurahman bin Muhammad bin Muhammad bin Muhammad bin Al-Hasan bin Jabir bin Muhammad bin Ibrahim bin Abdurahman bin Ibn Khaldun al-Hadrami. He was born in Tunis in AD 1332 into an upper-class Andalusian family of Arab descent. The family's ancestor was a Hadhrami who shared kinship with Waíl ibn Hujr, a companion of the Prophet. His family, which held many high offices in Al-Andalus, had emigrated to Tunisia after the fall of Seville in AD 1248. Although some of his family members had held political office in the Tunisian Hafsid dynasty, his father and grandfather later withdrew from political life and joined a mystical order. His brother, Yahya Khaldun, was also a historian who wrote a book on the Abdalwadid dynasty; Yahya was assassinated by a rival for being the official historiographer of the court.

In his autobiography, Ibn Khaldun traces his descent back to the time of the Prophet through an Arab tribe from Yemen, specifically the Hadhramaut, which came to the Iberian Peninsula in the 8th century, at the beginning of the Islamic conquest: "And our ancestry is from Hadhramaut, from the Arabs of Yemen, via Wa'il ibn Hujr also known as Hujr ibn 'Adi, from the best of the Arabs, well-known and respected." (p. 2429, Al-Waraq's edition).

His family's high rank enabled Ibn Khaldun to study with prominent teachers in Maghreb. He received a classical Islamic education, studying the Quran, which he memorized by heart, and Arabic linguistics which is the basis for understanding the Qur'an, hadith, sharia (law) and fiqh (jurisprudence). He received certification (ijazah) for all of those subjects. The mathematician and philosopher Al-Abili of Tlemcen introduced him to mathematics, logic and philosophy, and he studied especially the works of Averroes, Avicenna, Razi and Tusi. At the age of 17, Ibn Khaldūn lost both his parents to the Black Death, an intercontinental epidemic of the plague that hit Tunis in 1348–1349.

Following family tradition, he strove for a political career. In the face of a tumultuous political situation in North Africa, that required a high degree of skill in developing and dropping alliances prudently to avoid falling with the short-lived regimes of the time. Ibn Khaldūn's autobiography is the story of an adventure, in which he spends time in prison, reaches the highest offices and falls again into exile.

At the age of 20, he began his political career in the chancellery of the Tunisian ruler Ibn Tafrakin with the position of Kātib al-'Alāmah (seal-bearer), which consisted of writing in fine calligraphy the typical introductory notes of official documents. In 1352, Abū Ziad, the sultan of Constantine, marched on Tunis and defeated it. Ibn Khaldūn, in any case unhappy with his respected but politically meaningless position, followed his teacher Abili to Fez. There, the Marinid sultan, Abū Inan Fares I, appointed him as a writer of royal proclamations, but Ibn Khaldūn still schemed against his employer, which, in 1357, got the 25-year-old a 22-month prison sentence.

Upon the death of Abū Inan in 1358, Vizier al-Hasān ibn-Umar granted him freedom and reinstated him to his rank and offices. Ibn Khaldūn then schemed against Abū Inan's successor, Abū Salem Ibrahim III, with Abū Salem's exiled uncle, Abū Salem. When Abū Salem came to power, he gave Ibn Khaldūn a ministerial position, the first position to correspond with Ibn Khaldūn's ambitions.

The treatment that Ibn Khaldun received after the fall of Abū Salem through Ibn-Amar 'Abdullah, a friend of Ibn Khaldūn's, was not to his liking, as he received no significant official position. At the same time, Amar successfully prevented Ibn Khaldūn, whose political skills he knew well, from allying with the Abd al-Wadids in Tlemcen.

Ibn Khaldūn, therefore, decided to move to Granada. He could be sure of a positive welcome there since at Fez, he had helped the Sultan of Granada, the Nasrid Muhammad V, regain power from his temporary exile. In 1364, Muhammad entrusted him with a diplomatic mission to the king of Castile, Pedro the Cruel, to endorse a peace treaty. Ibn Khaldūn successfully carried out this mission and politely declined Pedro's offer to remain at his court and have his family's Spanish possessions returned to him.

In Granada, Ibn Khaldūn quickly came into competition with Muhammad's vizier, Ibn al-Khatib, who viewed the close relationship between Muhammad and Ibn Khaldūn with increasing mistrust. Ibn Khaldūn tried to shape the young Muhammad into his ideal of a wise ruler, an enterprise that Ibn al-Khatib thought foolish and a danger to peace in the country. History proved al-Khatib right, and at his instigation, Ibn Khaldūn was eventually sent back to North Africa. Al-

Khatib himself was later accused by Muhammad of having unorthodox philosophical views. Al-Khatib was murdered despite an attempt by Ibn Khaldūn to intercede on behalf of his old rival.

In his autobiography, Ibn Khaldūn tells little about his conflict with Ibn al-Khatib and the reasons for his departure.

Back in Ifriqiya, the Hafsid sultan of Bougie, Abū 'Abdallāh, who had been his companion in prison, received him with great enthusiasm and made Ibn Khaldūn his prime minister. Ibn Khaldūn carried out a daring mission to collect taxes among the local Berber tribes. After the death of Abū 'Abdallāh in 1366, Ibn Khaldūn changed sides once again and allied himself with the Sultan of Tlemcen, Abū l-Abbas. A few years later, he was taken prisoner by Abu Faris Abdul Aziz, who had defeated the sultan of Tlemcen and seized the throne.

Ibn Khaldūn then entered a monastic establishment and occupied himself with scholastic duties until 1370. In that year, he was sent for to Tlemcen by the new sultan. After the death of 'Abdu l-Azīz, he resided at Fez, enjoying the patronage and confidence of the regent.

Ibn Khaldūn's political skills and, above all, his good relationship with the wild Berber tribes were in high demand among the North African rulers, but he had begun to tire of politics and constantly switching allegiances. In 1375, he was sent by Abū Hammu, the 'Abdul Wadid Sultan of Tlemcen, on a mission to the Dawadida Arabs tribes of Biskra.

After his return to the West, Ibn Khaldūn sought refuge with one of the Berber tribes in the west of Algeria, in the town of Qalat Ibn Salama. He lived there for over three years under their protection, taking advantage of his seclusion to write the Muqaddimah "Prolegomena", the introduction to his planned history of the world.

In Ibn Salama, however, he lacked the necessary texts to complete the work. Therefore, in 1378, he returned to his native Tunis, which had meanwhile been conquered by Abū l-Abbas, who took Ibn Khaldūn back into his service. There, he devoted himself almost exclusively to his studies and completed his history of the world. His relationship with Abū l-Abbas remained strained, as the latter questioned his loyalty. That was brought into sharp contrast after Ibn Khaldūn presented him with a copy of the completed history that omitted the usual panegyric to the ruler. Under pretense of going on the Hajj to Mecca, something for which a Muslim ruler could not simply refuse permission, Ibn Khaldūn was able to leave Tunis and to sail to Alexandria.

Ibn Khaldun said of Egypt, "He who has not seen it does not know the power of Islam." While other Islamic regions had to cope with border wars and inner strife, Mamluk Egypt enjoyed prosperity and high culture. In 1384, the Egyptian Sultan, al-Malik udh-Dhahir Barquq, made Ibn Khaldun professor of the Qamhiyyah Madrasah and appointed him as the Grand qadi of the Maliki school of fiqh, widespread primarily in Western Africa. His efforts at reform encountered resistance, and within a year, he resigned his position. Also in 1384, a ship carrying Khaldun's wife and children sank off of Alexandria.

After his return from a pilgrimage to Mecca in May 1388, Ibn Khaldūn concentrated on teaching at various Cairo madrasas. At the

Mamluk court he fell from favor because during revolts against Barquq, he had, apparently under duress and along with other Cairo jurists, issued a fatwa against Barquq. Later relations with Barquq returned to normal, and he was once again named the Maliki qadi. Altogether, he was called six times to that high office, which, for various reasons, he never held long.

In 1401, under Barquq's successor (his son Faraj), Ibn Khaldūn took part in a military campaign against the Mongol conqueror, Timur, who besieged Damascus in 1400. Ibn Khaldūn cast doubt upon the viability of the venture and really wanted to stay in Egypt. His doubts were vindicated, as the young and inexperienced Faraj, concerned about a revolt in Egypt, left his army to its own devices in Syria and hurried home. Ibn Khaldūn remained at the besieged city for seven weeks, being lowered over the city wall by ropes to negotiate with Timur, in a historic series of meetings that he reported extensively in his autobiography.

Timur questioned him in detail about conditions in the lands of the Maghreb. At his request, Ibn Khaldūn even wrote a long report about it. As he recognized Timur's intentions, he did not hesitate, on his return to Egypt, to compose an equally-extensive report on the history of the Tatars, together with a character study of Timur, sending them to the Merinid rulers in Fez (Maghreb).

Ibn Khaldūn spent the next five years in Cairo completing his autobiography and his history of the world and acting as teacher and judge. Meanwhile, he was alleged to have joined an underground party, Rijal Hawa Rijal, whose reform-oriented ideals attracted the attention of local political authorities. The elderly Ibn Khaldun was placed under arrest. He died on 17 March 1406, one month after his sixth selection for the office of the Maliki qadi (Judge).

Modern historians in Europe and America are all praise for Ibn Khaldun. British historian Arnold J. Toynbee has called Ibn Khaldun's Muqaddimah "the greatest work of its kind." Ernest Gellner, once a professor of philosophy and logic at the London School of Economics, considered Khaldun's definition of government the best in the history of political theory.

Economist Paul Krugman described Ibn Khaldun as "a 14th-century philosopher who basically invented what we would now call the social sciences".

Scottish philosopher Robert Flint praised him strongly, "as a theorist of history he had no equal in any age or country until Vico appeared, more than three hundred years later. Plato, Aristotle, and Augustine were not his peers, and all others were unworthy of being even mentioned along with him".

71. Ismail ibn al-Ahmar

Abu l-Walid Ismail ibn Yusuf Ibn al-Ahmar

(Arabic: أبو الوليد إسماعيل بن يوسف ابن الأحمر),

(Granada 1324 – Fes 1407),

was a historian of the fourteenth century from Morocco. He was the official historian of the Marinid dynasty.

Contributions in Social Sciences

Ismail ibn al-Ahmar was the official historian of the Marinid dynasty. In addition to his works on history, he composed many works in poetry mainly in the praise of the Marinids and lamenting the feud with his royal cousin in Granada.

Books by Ibn al-Ahmar:

- Rawdat al-nisrin fi dawlat Bani Marin (written in 1404) A defense of Marinid policies, attacking the Abdalwadid of the Kingdom of Tlemcen.
- Nafha al-nisriniyya
- Nathir faraid al-djuman fi nazm fuhul al-azam, ed. Muhammad Ridwan al-Daya (Beyrouth 1967)

Biographical Summary

Ismail Ibn al-Ahmar was born in Granada in 1324. He was son of a Nasrid prince. He fled al-Andalus with his father as an infant and took refuge in Morocco where he was welcomed by the rival dynasty of the Marinids. He died in 1407 AD.

72. Taqi al-Din Muhammad al-Fasi

Taqi al-Din Muhammad ibn Ahmad al-Fasi

(Arabic: تقي الدين أبي الطيب محمد بن أحمد الفاسي),

(Mecca 8 September 1373 – Mecca 6 July 1429),

was a hafith, faqih, historian, and Maliki qadi (judge) in Mecca. He is best known for his works on the history of Mecca and its rulers and notable natives, around 18 works. He also wrote on the genealogies of some Arab tribes of Tihamah.

Contributions in Social Sciences

Following are some of the works by Taqi al-Din Muhammad al-Fasi.

- Al-'Iqd al-thamīn fī tārīkh al-Balad al-Amīn (العقد الثمين فى تاريخ البلد الأمين): His largest and most important work, and probably the largest in the field of Meccan history, where he compiled the biographies of Meccans from the early days of Islam up until his time.
- Shifā' al-gharām bi-akhbār al-Balad al-Ḥarām (شفاء الغرام بأخبار البلد الحرام)
- Al-Muqni' min akhbār al-mulūk wa-al-khulafā' wa-wulāt Makkah al-shurafā (المقنع من أخبار الملوك والخلفاء وولاة مكة الشرفاء)
- Al-Zuhūr al-muqtaṭafah min tārīkh Makkah al-Musharrafah (الزهور المقتطفة من تاريخ مكة المشرفة)
- Dhayl al-taqīīd fī rūāh al-sunan wa al-masānīd (ذيل التقييد بمعرفة رواة السنن والمسانيد)

Biographical Summary

He was born on Thursday, 8 September 1373 in Mecca, Hejaz, Arabian Peninsula. He spent part of his early life in Medina. He eventually returned to Mecca where he took knowledge from its scholars. His family claimed descent from the Islamic prophet Muhammad through his grandson, Hasan ibn Ali. He was a teacher of Maliki fiqh at the Ghiyathiyyah Madrasah in Makkah, which was considered one of the best Islamic institutions in the country and was funded by the Sultan of Bengal Ghiyasuddin Azam Shah. He went blind four years before his death in 1425 AD. He died on Wednesday 6 July 1429 at the age of 55, in Mecca, Hejaz, Arabian Peninsula.

73. Ahmad Ibn Arabshah

Abu Muhammad Shihab al-Din Ahmad ibn Muhammad ibn Abd Allah ibn Ibrahim also known as Muhammad ibn Arabshah

(1389–1450),

was a writer and traveler who lived under the reign of Timur (1370–1405).

Contributions in Social Sciences

Following are some of the works by ibn Arabshah:

- Aja'ib al-Maqdur fi Nawa'ib al-Taymur (The Wonders of Destiny of the Ravages of Timur), which he finished in Damascus on 12 August 1435. This book was translated and printed first time in Latin; Ahmedis Arabsiadae Vitae & rerum gestarum Timuri, qui vulgo Tamerlanes dicitur, historia. Lugduni Batavorum, ex typographia Elseviriana, 1636.
- al-Ta'lif al-tahir fi shiyam al-Malik al-Zahir (Life of Zahir)
- Fakihat al-Khulafa' wa Mufakahat al-Zurafa'
- Jami' al-Hikayat; translated from Persian to Turkish.
- al-'Aqd al-Farid fi al-Tawhid
- Ghurrat al-Siyar fi Duwal al-Turk wa al-Tatar
- Muntaha al-Adab fi Lughat al-Turk wa al-Ajam wa al-'Arab

Biographical Summary

Ibn Arabshah was born in 1389 AD and grew up in Damascus. Later when Timur invaded Syria, he moved to Samarkand and later to Transoxiana. He later moved to Edirne and worked in the court of

Sultan Mehmed I translating Arabic books to Turkish and Persian. He later returned to Damascus after having been absent from the city for 23 years. Later he moved to Egypt and died there in 1450 AD.

The famous Muslim scholar, Abd al-Wahhab ibn Arabshah, was his son.

74. Al-Maqrizi

Taqī al-Dīn Abū al-'Abbās Aḥmad ibn 'Alī ibn 'Abd al-Qādir ibn Muḥammad al-Maqrīzī

(Arabic: تقي الدين أحمد بن علي بن عبد القادر بن محمد المقريزي),

(1364–1442),

was a historian from Egypt during the Mamluk era. He is known for his interest in the Fatimid dynasty and its role in Egyptian history.

Contributions in Social Sciences

Al-Maqrizi's works exceed 200, and mostly are concerned with Egypt. However, it is stated that his books are mostly compilations and he does not always give references to his sources.

- Muqaffa, first sixteen-volumes of an Egyptian biographic encyclopedia arranged in alphabetic order. The Egyptian historian, al-Sakhawi, estimated that the complete work would require eighty volumes. Three autograph volumes exist in manuscript in Leiden and one in Paris.
- Al-Mawāʻiẓ wa-al-Iʻtibār bi-Dhikr al-Khiṭaṭ wa-al-āthār (Arabic, 2 vols.) Būlāq al-Qāhirah: Dār al-Ṭibāʻah al-Miṣrīyah.
- Ittiʻāz al-Ḥunafā' bi-Akhbār al-A'immah al-Fāṭimīyīn al-Khulafā'
- Kitāb al-Khiṭaṭ al-Maqrīzīyah
- Kitāb al-Sulūk li-Maʻrifat Duwal al-Mulūk
- History of the Fatimites; extract published by J.G.L. Kosegarten in Chrestomathia (Leipzig, 1828), pp. 115–123;

- History of the Ayyubit and Mameluke Rulers; French translation by Etienne Marc Quatremère (2 vols., Paris, 1837–1845).

Following are modern translation editions of some smaller works of Al-Maqrizi:

- Mahomeddan Coinage (ed. O. G. Tychsen, Rostock, 1797; French translation by Silvestre de Sacy, Paris, 1797)
- Arab Weights and Measures (ed. Tychsen, Rostock, 1800)
- Arabian Tribes that migrated to Egypt (ed. F. Wüstenfeld, Göttingen, 1847)
- Account of Hadhramaut (ed. P.B. Noskowyj, Bonn, 1866)
- Strife between the Bani Umayya and the Bani Hashim (ed. G. Vos, Leiden, 1888)
- Historia Regum Islamiticorum in Abyssinia (ed. and Latin trans. F. T. Rink, Leiden, 1790).

Some books on the works of Al-Maqrizi:

- Al Mawaiz wa al-'i'tibar bi dhikr al-khitat wa al-'athar (about the planning of Cairo and its monuments)
- A. R. Guest, p. 103ff: A list of Writers, Books and other Authorities mentioned by El Maqrisi in his Khitat
- Al Selouk Leme'refatt Dewall al-Melouk (about Mamluk history in Egypt)
- Ette'aaz al-honafa be Akhbaar al-A'emma Al Fatemeyyeen Al Kholafaa (about the Fatimid state)
- Al Bayaan wal E'raab Amma Be Ard Misr min al A'raab (about the Arab Tribes in Egypt)

- Eghathatt Al Omma be Kashf Al Ghomma (about the famines that took place in Egypt)
- Al Muqaffa (biographies of princes and prominent personality of his time)
- *Maqrīzī, Aḥmad ibn ʿAlī (1824).* Hamaker, *Hendrik Arent (ed.).* Takyoddini Ahmedis al-Makrizii Narratio de Expeditionibus a Graecis Francisque Adversus Dimyatham, AB A. C. 708 AD 1221 Susceptis *(in Arabic and Latin).* Amstelodami*: Pieper & Ipenbuur.*
- Kosegarten, J. G. L. *(1828).* Chrestomathia Arabica ex codicibus manuscriptis Parisiensibus, Gothanis et Berolinensibus collecta atque tum adscriptis vocalibus *(in Latin). Lipsiae: Sumtu F. C. G. Vogelii.* (pp. 115 –123: Al-Maqrizi, an extract of History of the Fatimites.)
- *Al-Maqrizi (1840).* Histoire des sultans mamlouks, de l'Égypte, écrite en arabe *(in French and Latin). Vol. 1, part 1. Translator:* Étienne Marc Quatremère. *Paris: Oriental Translation Fund of Great Britain and Ireland.*
- *Al-Maqrizi (1845).* Histoire des sultans mamlouks, de l'Égypte, écrite en arabe *(in French and Latin). Vol. 1, part 2. Translator:* Étienne Marc Quatremère. *Paris: Oriental Translation Fund of Great Britain and Ireland.*
- *Al-Maqrizi (1845).* Histoire des sultans mamlouks, de l'Égypte, écrite en arabe *(in French and Latin). Vol. 2. Translator:* Étienne Marc Quatremère. *Paris: Oriental Translation Fund of Great Britain and Ireland.*

- *Al-Maqrizi (1853).* Kitb al-mawi wa-al-itibr bi-dhikr al-khia wa-al-thr : yakhtau dhlika bi-akhbr iqlm Mir wa-al-Nl wa-dhikr al-Qhirah wa-m yataallaqu bi-h wa-bi-iqlmih *(in Arabic). Vol. 1. al-Qhirah: Maktabat al-Thaqfah al-Dnyah.*
- *Al-Maqrizi (1853).* Kitb al-mawi wa-al-itibr bi-dhikr al-khia wa-al-thr : yakhtau dhlika bi-akhbr iqlm Mir wa-al-Nl wa-dhikr al-Qhirah wa-m yataallaqu bi-h wa-bi-iqlmih *(in Arabic). Vol. 2. al-Qhirah: Maktabat al-Thaqfah al-Dnyah.*
- A. R. Guest, 1902, in Journal of the Royal Asiatic Society, pp. 103–125: "A List of Writers, Books and other Authorities mentioned by El Maqrizi in his Khitat" (Notes on the 1853-edition)
- *Al-Maqrizi (1895).* Mémoires publiés par les membres de la Mission archéologique Française au Caire: Description topographique et historique de l'Égypte *(in French). Vol. 17. Translator:* Urbain Bouriant. *Cairo: Mémoires publiés par les membres de la Mission archéologique.*
- *Al-Maqrizi (1948). Jamāl al-Dīn al-Shayyāl (ed.). Itti'āz al-Ḥunafā' bi-Akhbār al-A'immah al-Fāṭimīyīn al-Khulafā'.* Cairo*: Dār al-Fikr al-'Arabī.*
- *Al-Maqrizi (1908) [1906]. Kitāb al-Khiṭaṭ al-Maqrīzīyah. Vol. 4. Cairo: Al-Nil Press.*
- *Al-Maqrizi (1956). Muḥammad Muṣṭafā Ziada (al-Ziyādah) (ed.). Kitāb al-Sulūk li-Ma'rifat Duwal al-Mulūk. Vol. 3. Cairo: Lajnat al-Ta'līf.*

- Alternative: *Al-Maqrizi (1895).* Mémoires publiés par les membres de la Mission archéologique Française au Caire: Description topographique et historique de l'Égypte *(in French). Vol. 17. Translator:* Urbain Bouriant. *Cairo: Mémoires publiés par les membres de la Mission archéologique.*
- French translation by Urbain Bouriant as Description topographique et historique de l'Égypte (Paris, 1895–1900; compare A. R. Guest"
- A List of Writers, Books and other Authorities mentioned by El Maqrizi in his Khitat," in Journal of the Royal Asiatic Society, 1902, pp. 103–125).

Biographical Summary

A direct student of Ibn Khaldun, Al-Maqrīzī was born in Cairo and spent most of his life in Egypt. When he presents himself in his books he usually stops at the 10th forefather although he confessed to some of his close friends that he can trace his ancestry to Al-Mu'izz li-Dīn Allāh – first Fatimid caliph in Egypt and the founder of al-Qahirah – and even to Ali ibn Abi Talib.

He was trained in the Hanafite school of law. Later, he switched to the Shafi'ite school and finally to the Zahirite school. Maqrizi studied theology under one of the primary masterminds behind the Zahiri Revolt, and his vocal support and sympathy with that revolt against the Mamluks likely cost him higher positions with the Mamluk regime. The name Maqrizi was an attribution to a quarter of the city of Baalbek, from where his paternal grandparents hailed. Maqrizi confessed to his contemporaries that he believed that he was related to the Fatimids through the son of al-Muizz. Ibn Hajar preserves the most memorable account: his father, as they entered the al-Hakim Mosque one day, told him "My son, you are entering the mosque of your ancestor." However, his father also instructed al-Maqrizi not to reveal this information to anyone he could not trust.

In 1385, he went on the Hajj.

For some time, he was secretary in a government office, and in 1399 became inspector of markets for Cairo and northern Egypt. This post he soon gave up to become: a preacher at the Mosque of 'Amr ibn al 'As, president of the al-Hakim Mosque, and a lecturer on Hadsith.

In 1408, he went to Damascus to become inspector of the Qalanisryya and lecturer. Later, he retired into private life at Cairo.

In 1430, he again went on Hajj with his family and travelled for some five years. He died in 1442 AD.

Concluding Remarks

We have presented 74 scientists in the Social Sciences from the part 1 (AD 610 to 1400) of the Islamic Era (AD 610 to 1922). All of them except one are Muslims, an expression of the fact that the era was entirely dominated by the Muslims in all domains of social sciences.

What are the reasons that Muslims excelled in sciences par excellence. The series makes it explicit that there is a natural affinity between Islam and science because Quran exhorts its readers to a scientific outlook in life by urging them to observe the nature and the universe around them to get to know them. That is the solidly open path to appreciate the truth in Quran and to approach closer to its Speaker.

Following are some examples of world-leading accomplishments of the social scientists of the Islamic Era that had led to renaissance in Europe, as the European social scientists stood on the shoulders of such giants.

- The Umayyad and Abbasid Caliphs were in continuous international diplomatic negotiations on matters such as peace treaties, the exchange of prisoners of war, and payment of ransoms and tributes. Al-Shaybani wrote Introduction to the Law of Nations at the end of the 8th century. These developed a number of modern topics in international law, thus founding the modern international laws and political science in international affairs. Examples of these topics include the following: the law of treaties; the treatment of diplomats; laws about hostages, refugees, prisoners of war, and the right of asylum; laws of war and peace regarding the conduct on the battlefield, protection of women, children and non-combatant civilians, and contracts across the lines of battle; laws of human values such as the use of poisonous weapons and devastation of enemy territory.

- Al-Baladhuri wrote the treatise "Book of the Conquests of Lands" laying down the foundations of political science, elaborating such topics as the terms made with the residents of the conquered territories. It tells of the conquests of the Arabs from the 7th century, covering conquests of lands from Arabia west to Egypt, North Africa, and Spain; and east from Arabia to Iraq, Iran, and Sind.
- Al-Farabi was a philosopher and music theorist. He is regarded as "Father of Neoplatonism", and "Founder of Political Philosophy".
- Ibn Fadlan pioneered the approach to combine diplomatic service with historic, geographic and cultural observations. He pioneered the use of his travel accounts connected with his diplomatic assignments. His writings provide unique accounts of Volga Vikings, while he served as a member of the embassy of the Abbasid caliph in Volga Bulgars.
- Al-Masʿūdī is the father of Social History, with his celebrated magnum opus "Murūj al-Dhahab wa-Ma'ādin al-Jawhar" (The Meadows of Gold and Mines of Gems) combining universal

history with scientific geography and social commentary and biography.

- Al-Maqdisī was the father of geography, as he was the first to desire, conceive and treat geography as an "original science". His treatise "Aḥsan al-taqāsīm fī maʿrifat al-aqālīm" (The Best Divisions in the Knowledge of the Regions) establishes this approach for the first time.
- Ibn Rushd is the father of rationalism, and his writings triggered a philosophical movement in Europe, called Averroism. In medicine, Ibn Rushd proposed a new theory of stroke, described the signs and symptoms of Parkinson's disease for the first time, and was the first to identify the retina as the part of the eye responsible for sensing light.
- Ibn Battuta established world's most extensive exploration record, traveling continuously over a period of thirty years from 1325 to 1354. He visited most of North Africa, the Middle East, East Africa, Central Asia, South Asia, Southeast Asia, China, the Iberian Peninsula, and West Africa. Ibn Battuta travelled more than any other explorer in pre-modern history, totaling around 117,000 km - surpassing Zheng He with about 50,000 km and Marco Polo with 24,000 km.
- Ibn Khaldūn is the founder of the proto-disciplines that would become historiography, sociology, economics, and demography.
- Taqi al-Din Muhammad al-Fasi founded the proto-discipline that remains 'unnamed' till today. He dedicated himself to the

history of just one city (Mecca), its rulers, its notable natives, and the genealogies of its tribes. We will name this discipline as "Baladology".

This series on "scientists of the Islamic Era" demonstrates that Muslim scientists are the giants on whose shoulders present-day sciences are built. The series includes but only a few from the Muslim science community in Islamic Era. There are innumerable others, and many have been lost to oblivion. There is a wealth of "science" buried in that community and it remains to be extracted from the archives. Researchers will no doubt make further discoveries. Subsequent editions of this book would expand the set of scientists included, as well as additional details about those already covered.

There is at least a three-fold purpose to this book series. One is to invite the world science community, in a manner of civilizational dialogue, to celebrate the science giants that Islamic Era has contributed to the growth of science and technology at its foundational level, as well as at the level of expanding its frontiers. Another is to remind the Muslims of their love for "science" which every man and woman must acquire; not for worldly dominance, but for a better humanity in a better world. One other objective is to join hands with the rest of humanity by satisfying the upwelling desire of the youth to know the truth about Muslim civilization and the excellence of their pursuit for wholistic knowledge: scientific, humanitarian, cultural, civilizational, and spiritual.

It is time for the world to move ahead of the historical biases, religious prejudices, cultural entanglements, and hegemonic aspirations. All people, together, constitute our humanity, and we hold this truth as self-evident that all humans are created with equal value. So, let us all join hands to bring science in the service of making every day a wonderful day in every neighborhood of our planet.

www.ingramcontent.com/pod-product-compliance
Lightning Source LLC
Chambersburg PA
CBHW070053030426
42335CB00016B/1868
* 9 7 8 1 9 5 6 6 0 1 2 0 6 *